AF595468

Bifurcation and Chaos in Fractional-Order Systems

Bifurcation and Chaos in Fractional-Order Systems

Editors

Marius-F. Danca
Guanrong Chen

MDPI • Basel • Beijing • Wuhan • Barcelona • Belgrade • Manchester • Tokyo • Cluj • Tianjin

Editors
Marius-F. Danca
Romanian Instituite of Science and Technology
Romania

Guanrong Chen
Department of Electrical Engineering,
City University of Hong Kong
China

Editorial Office
MDPI
St. Alban-Anlage 66
4052 Basel, Switzerland

This is a reprint of articles from the Special Issue published online in the open access journal *Symmetry* (ISSN 2073-8994) (available at: https://www.mdpi.com/journal/symmetry/special_issues/Bifurcation_Chaos_Fractional-Order_Systems).

For citation purposes, cite each article independently as indicated on the article page online and as indicated below:

LastName, A.A.; LastName, B.B.; LastName, C.C. Article Title. *Journal Name* **Year**, *Volume Number*, Page Range.

ISBN 978-3-0365-0092-8 (Hbk)
ISBN 978-3-0365-0093-5 (PDF)

Cover image courtesy of Marius-F. Danca.

Contents

About the Editors

Marius-F. Danca received an MSc degree in Mathematics in 1980 from Babes-Bolyai University of Cluj-Napoca, Romania, Faculty of Mathematics; an MSc in Electrotechnics in 1986 from the Technical University of Cluj-Napoca, Romania, Faculty of Electrotechnics; a PhD degree in Automation in 1997 from the Technical University of Cluj-Napoca, Romania, Faculty of Automation and Computer Science; and a PhD degree in Mathematics in 2002 from Babes-Bolyai University of Cluj-Napoca, Romania, Faculty of Mathematics.

Professor Marius-F. Danca was Guest Associate Editor at Discrete and Continuous Dynamical Systems—Series S (DCDS-S) (2007–2017), Associate Editor at Dynamics of Continuous, Discrete & Impulsive Systems—Series B (DCDIS-B) (2006–present), Associate Editor at Journal of Nonlinear Systems and Applications (JNSA) (2009–present), and Guest Editor of the Special Issue "Research Frontier in Chaos Theory and Complex Networks" (2018).

Professor Marius-F. Danca has 85 ISI papers, an h-index of 18, and more than 200 publications in Romanian and foreign dissemination journals, and was a reviewer for more than 40 ISI journals (at Springer, Elsevier, Taylor & Francis, AIMS, Willey, IET, Hindawi, World Scientific, MDPI, AIP, John Willey & Sons, Hacettepe, IOP, and Sage).

He is the author of two books, entitled "Functia logistica: dinamica, bifurcatie si haos" Seria MATEMATICA APLICATA SI INDUSTRIALA 7, Editura Universitatii din Pitesti, 2001; and "Sisteme dinamice discontinue" Seria MATEMATICA APLICATA SI INDUSTRIALA 14, Editura Universitatii din Pitesti, 2004 (in Romanian); and one chapter in "Complex Systems and Networks: Dynamics, Controls and Applications" in Understanding Complex Systems, Springer 2016.

Guanrong Chen received an MSc degree in Computer Science from Sun Yat-sen University, Guangzhou, China, in 1981 and a PhD degree in Applied Mathematics from Texas A&M University, USA, in 1987. Since 2000, he has been a Chair Professor and the founding director of the "Centre for Chaos and Complex Networks" at the City University of Hong Kong.

Professor Chen was elected a Fellow of the IEEE in 1997, awarded the 2011 Euler Gold Medal from Russia, and conferred Honorary Doctor Degrees by the Saint Petersburg State University, Russia, in 2011 and by the University of Normandy, France, in 2014. He is a Member of the Academy of Europe (since 2014) and a Fellow of The World Academy of Sciences (since 2015).

Professor Chen's research interests are in the fields of complex networks, nonlinear dynamics, and control systems. He is a Highly Cited Researcher in Engineering (since 2009), according to Thomson Reuters.

Preface to "Bifurcation and Chaos in Fractional-Order Systems"

The concept of fractional-order differentiation first emerged in 1965 regarding a historical correspondence between the Marquise de L'Hospital and the mathematician Leibnitz. In the sequel, famous mathematicians such as Euler, Laplace, Abel, Liouville, and Riemann further developed fundamental technical details. It was realized recently that many scientific phenomena with complex dynamics cannot be well modeled by differential equations using integer-order derivatives. Fractional calculus (calculus of non-integer order) became a rapidly developing topic in science and engineering, which has been attracting a great deal of attention recently in the academic and industrial world. While exponential laws represent a classical approach to studying dynamical systems, there are systems where faster or slower dynamics are better and more accurately described by Mittag–Leffler functions. As a result, there has been an increasing interest to merge the mathematical fundamentals of fractional calculus into scientific and engineering applications as an interdisciplinary approach, which has started to demonstrate some advantages over conventional integer-order differential systems. Although the last decade had witnessed significant development in this research area, many theoretical and technical problems remain to be further explored, including particularly chaotic fractional-order systems. On the other hand, finding hidden attractors in continuous-time and discrete-time chaotic fractional-order systems represents a new trend of research, as an exciting and challenging direction in the fields of complex dynamics. Of particular interest are those systems with symmetry, in which bifurcations can lead to a family of conjugate attractors, all connected by symmetry. Therefore, this research direction of bifurcation and chaos in fractional-order dynamical systems opens up a corpus of opportunities with encouraging promises in such scientific fields as complex dynamics, systems and networks, and signal processing, to name just a few.

The book consists of seven contributed papers written by active leading experts on various topics.

The first paper, entitled "Fractional Dynamics in Soccer Leagues", addresses the dynamics of four European soccer teams over the 2018–2019 season, with modeling based on fractional calculus and power law. The new model embeds implicit details such as the behavior of players and coaches, strategical and tactical maneuvers during the matches, errors of referees, and a multitude of other effects. Two approaches are taken to evaluate the teams' progress along the league by models fitting real-world data and to analyze statistical information by using entropy.

The second paper, entitled "Puu System of Fractional Order and Its Chaos Suppression", studies the fractional-order variant of Puu's system and compares it with the integer-order counterpart. Also, an impulsive chaos control algorithm is applied to suppress chaos in the system.

The third paper, entitled "Dynamical Properties of Fractional-Order Memristor", investigates the properties of a fractional-order memristor, revealing the influences of parameters such as the fraction order, frequency, switch resistor ratio, average mobility on the system dynamics, and so on. The fractional-order memristor is implemented by circuits.

The forth paper, entitled "Fractional Levy Stable and Maximum Lyapunov Exponent for Wind Speed Prediction", proposes a wind speed prediction method based on the maximum Lyapunov exponent and the fractional Levy stable motion in an iterative prediction model. Theoretical analysis and numerical simulations are performed with comparisons, showing some advantages of the new method.

The fifth paper, entitled "On Two-Dimensional Fractional Chaotic Maps with Symmetries", discusses two new two-dimensional chaotic maps with closed curve fixed points. It analyzes the chaotic behavior of the two maps by the 0–1 test and explores it numerically using Lyapunov exponents and bifurcation diagrams, showing that chaos exists in both fractional maps and that the fractional-order map has coexisting attractors.

The sixth paper, entitled "Bifurcations, Hidden Chaos and Control in Fractional Maps", introduces, based on discrete fractional calculus, simple two-dimensional and three-dimensional fractional maps; both are chaotic and have a unique equilibrium point, with coexisting attractors. Moreover, control schemes are introduced to stabilize the chaotic trajectories of the two systems.

The seventh paper, entitled "Generalized Bessel Polynomial for Multi-Order Fractional Differential Equations", defines a simple but effective method for approximating solutions of multi-order fractional differential equations relying on the Caputo fractional derivative under some conditions. Basis functions used are generalized Bessel polynomials satisfying many properties shared by the classical orthogonal polynomials of Hermit, Laguerre, and Jacobi. The new method has good performance in accuracy and simplicity. Some practical test problems with symmetries are used to verify the proposed technique with comparisons.

We thank all the authors for their excellent research and fine contributions to this edited book.

Marius-F. Danca, Guanrong Chen

Editors

Article

Fractional Dynamics in Soccer Leagues

António M. Lopes [1,*] and Jose A. Tenreiro Machado [2]

[1] UISPA–LAETA/INEGI, Faculty of Engineering, University of Porto, Rua Dr. Roberto Frias, 4200-465 Porto, Portugal

[2] Institute of Engineering, Polytechnic of Porto, Department of Electrical Engineering, Rua Dr. António Bernardino de Almeida, 431, 4249-015 Porto, Portugal; jtm@isep.ipp.pt

* Correspondence: aml@fe.up.pt

Received: 27 January 2020; Accepted: 16 February 2020; Published: 1 March 2020

Abstract: This paper addresses the dynamics of four European soccer teams over the season 2018–2019. The modeling perspective adopts the concepts of fractional calculus and power law. The proposed model embeds implicitly details such as the behavior of players and coaches, strategical and tactical maneuvers during the matches, errors of referees and a multitude of other effects. The scale of observation focuses the teams' behavior at each round. Two approaches are considered, namely the evaluation of the team progress along the league by a variety of heuristic models fitting real-world data, and the analysis of statistical information by means of entropy. The best models are also adopted for predicting the future results and their performance compared with the real outcome. The computational and mathematical modeling lead to results that are analyzed and interpreted in the light of fractional dynamics. The emergence of patterns both with the heuristic modeling and the entropy analysis highlight similarities in different national leagues and point towards some underlying complex dynamics.

Keywords: Fractional dynamics; Power law behavior; Complex systems; Soccer

---

1. Introduction

Soccer is the most popular sport in Europe [1,2]. The game is played by two teams of 11 players, on a rectangular field with a goal placed at each end. The objective of the game is to score by getting a spherical ball into the opposing goal. Each team includes 10 field players, that can maneuver the ball using any part of the body except hands and arms, and one goalkeeper, who is allowed to touch the ball with the whole body, as long as they stay in their penalty area. Otherwise, the rules of the field players apply. The match has two periods of 45 minutes each. The winning team is the one that scores more goals by the end of the match.

In most European countries, soccer competitions are organized hierarchically in leagues composed by groups of teams. At the end of each season, a promotion and relegation system decides which teams move up and down into the hierarchy. In a given league and season each pair of teams plays to matches, so that the visited and visitor interchange place. All teams start with zero points and, at every round, one {victory, draw, defeat} is worth $\{3, 1, 0\}$ points. By the end of the last round, the team that accumulated more points is the champion [3].

This paper studies the dynamical performance of soccer teams in a given league. The modeling perspective adopts the concepts of fractional calculus [4,5] and power law (PL) [6]. These tools have been used to model dynamical systems with memory in mechanics [7], electromagnetism [8], biology [9], sports [10], economy [11], and others [12]. The proposed approach embeds implicitly details such as the behavior of players and coaches, strategical and tactical maneuvers during the matches, errors of referees and a multitude of other effects. The scale of observation addresses the teams behavior in the perspective of their classification along the league. Data characterizing the year

2018–2019 and the four leagues, namely the Spanish 'La Liga', English 'Premiership', Italian 'Serie A' and French 'Ligue 1', are processed and discussed. The computational and mathematical modeling leads to the emergence of patterns that are analyzed and interpreted in the light of fractional dynamics.

In spite of its social and economical importance, the topic of soccer dynamics has been the subject of only a restricted number of studies. In fact, we can consider the study of the dynamic effects at different levels, namely about the evolution of each player along his career, of the evolution of the players within a given match, or the progress of the classification of a group of teams along a given league. Couceiro et al. [13] characterized the predictability and stability levels of players during a soccer match by means of fractional calculus and entropy measures. Silva et al. [14] investigated how different entropy measures can be applied to assess the performance variability and to uncover the interactions underlying the players and teams' performance. Machado and Lopes [15] adopted distinct dissimilarity measures and multidimensional scaling to study the behavior of teams competing within soccer national leagues and related the generated loci with the teams' performance over time. Neuman et al. [16] measured the organization associated with the behavior of a soccer team through the Tsallis entropy of ball passes between the players. They found that the teams' positions at the end of one season were correlated with the teams entropy. Moreover, the entropy score could be used for predicting of the teams' final positions. Lopes and Machado [17] studied the dynamics of national soccer leagues using information and fractional calculus tools. In their approach, an entire soccer league season was treated as a complex system with a state observable at the time of rounds, consisting of the goals scored by the teams. The system behavior was visualized in 3-D maps generated by multidimensional scaling and interpreted based on the emerging clusters.

Predicting the outcome of soccer matches has been investigated since at least the 1960s, due to its interests for the general public, clubs, advertising companies, media, professional odds setters, and researchers [18]. Various statistical techniques have been used, including Poisson models [19], Bayesian schemes [20], rating systems [21], and machine learning methods [22,23], such as kernel-based relational learning [24], among others [25,26]. Advanced machine learning techniques [27,28] may be of relevance and represent an alternative to statistical or analytical approaches. However, dynamical phenomena involving complex memory effects need to be analyzed under the light of modeling tools to better understand phenomena embedded in the data series. Therefore, a synergistic strategy encompassing studies of distinct nature seems the best to consider at the moment of writing this paper.

This paper addresses the dynamics exhibited by several leagues having in mind the evolution of the teams along one season. The adoption of heuristic models, fitting real-world data, and the entropy analysis of statistical information, give a new perspective to a complex system that has been scarcely studied in the scientific literature.

Bearing these ideas in mind, this paper is organized as follows. Section 2 models the behavior of the teams in four top European soccer leagues by means of different functions. Section 3 analyzes the leagues in the perspective of the entropy of the spatio-temporal patterns exhibited by distinct alternative models. Section 4 uses the models for predicting future results and assesses their accuracy. Finally, Section 5 outlines the conclusions.

2. Modeling the Teams' Dynamics

Let us consider N teams competing in a league for one season. Therefore, the league has $R = 2(N-1)$ rounds, and each team plays $N-1$ matches at home and $N-1$ matches away.

Let us denote by $x_i(k), i = 1, \ldots, N, 0 \leq k \leq k_r$, the teams' positions from the beginning of the season up to the round $k_r = 3, \ldots, R$. The lower limit $k_r = 3$ is adopted to yield data-series with a minimum number of points for processing. Therefore, the signals $x_i(k)$ evolve in discrete time and one-dimensional space, and can be seen as the output of a complex system.

We use the nonlinear least-squares [29,30] to test the behavior of the series $x_i(k)$ for six fitting hypotheses, namely shifted power (SP), quadratic (Qu), Hill (Hi), vapor pressure (VP), power law (PL) and Hoerl (Ho) models, given by:

$$SP: \ \hat{x}_i^{SP}(k) = a_i(k_r) \cdot |k - b_i(k_r)|^{c_i(k_r)}, \tag{1a}$$

$$Qu: \ \hat{x}_i^{Qu}(k) = a_i(k_r) + b_i(k_r) \cdot k + c_i(k_r) \cdot k^2, \tag{1b}$$

$$Hi: \ \hat{x}_i^{Hi}(k) = \frac{a_i(k_r) \cdot k^{b_i(k_r)}}{c_i(k_r)^{b_i(k_r)} + k^{b_i(k_r)}}, \tag{1c}$$

$$VP: \ \hat{x}_i^{VP}(k) = e^{\frac{a_i(k_r)+b_i(k_r)}{c_i(k_r)\cdot \ln(k)}}, \tag{1d}$$

$$PL: \ \hat{x}_i^{PL}(k) = a_i(k_r) \cdot k^{b_i(k_r)}, \tag{1e}$$

$$Ho: \ \hat{x}_i^{Ho}(k) = a_i(k_r) \cdot b_i(k_r)^k \cdot k^{c_i(k_r)}, \tag{1f}$$

where $\hat{x}_i$ denotes the approximated values, k represents time and $\{a_i(k_r), b_i(k_r), c_i(k_r)\} \in \mathbb{R}$ are the models' parameters. Naturally, for each model, the parameters vary with time, that is, they depend on k_r.

We can adopt other fitting models, eventually with more parameters, that adjust better to some particular series $x_i(k)$. However, only simple analytical expressions, requiring a limited set of parameters, are considered [31], otherwise the interpretation of the parameters becomes unclear. Moreover, loosely speaking, with exception of Qu, these heuristic models reflect somehow fractional characteristics, embodied in their structures by the non-integer exponents.

For assessing the accuracy of the models (1a)–(1f) we adopt the time varying errors E_i and $E^{\dagger}$ given by:

$$E_i(k_r) = x_i(k_r) - \hat{x}_i(k_r), \ i = 1, \cdots, N, \ k_r = 3, \cdots, R, \tag{2a}$$

$$E^{\dagger}(k_r) = \sqrt{\frac{1}{N}\sum_{i=1}^{N} [E_i(k_r)]^2}, \ k_r = 3, \cdots, R, \tag{2b}$$

where $N = 20$ and $R = 38$ correspond to the number of teams and rounds in the 'La Liga', 'Premiership', 'Serie A' and 'Ligue 1'. Therefore, E_i gives information for team i, while $E^{\dagger}$ highlights the fitting error for all teams involved in the league.

Figure 1 illustrates the error $E_1(k_r)$ for the 2018–2019 champions of 'La Liga' and "Premiership", namely the FC Barcelona and Manchester City, and the models (1a)–(1f). Figure 2 depicts $E^{\dagger}(k_r)$ for 'La Liga' and 'Premiership'. Obviously, this error increases with the number of data points. Moreover, we verify that, with the exception of the Hi (1c), all other models approximate well the data, demonstrating the adequacy of the fitting functions. The other teams and leagues yield charts of the same type and, therefore, are omitted herein. For a more assertive comparison, we calculate the mean and the standard deviation, $\mu = \frac{1}{36}\sum_{k_r=3}^{38} E^{\dagger}(k_r)$ and $\sigma = \sqrt{\frac{1}{36}\sum_{k_r=3}^{38}[E^{\dagger}(k_r) - \mu]^2}$, of the $E^{\dagger}(k_r)$ series for the models (1a)–(1f). Table 1 summarizes the corresponding values, highlighting that the Ho (1f) is the best three-parameter model approximation, and that the PL (1e) represents a good alternative, when having in mind the advantage of having just two parameters.

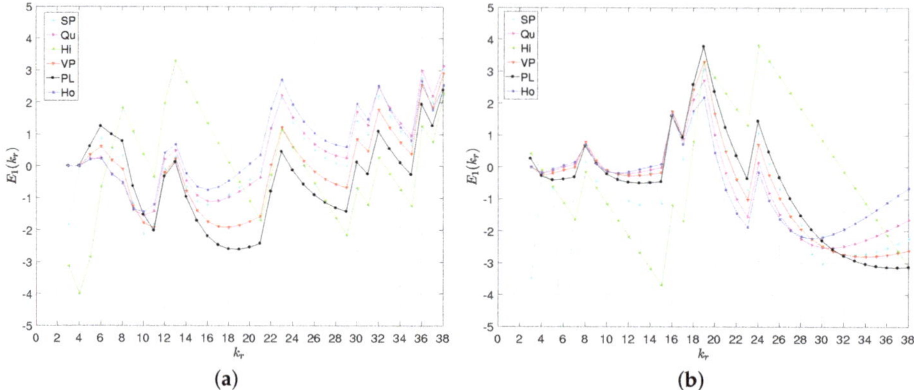

Figure 1. The error $E_1(k_r)$, $k_r = 3, \ldots, 38$, for the models (1a)–(1f), during the season 2018–2019: (**a**) FC Barcelona; (**b**) Manchester City.

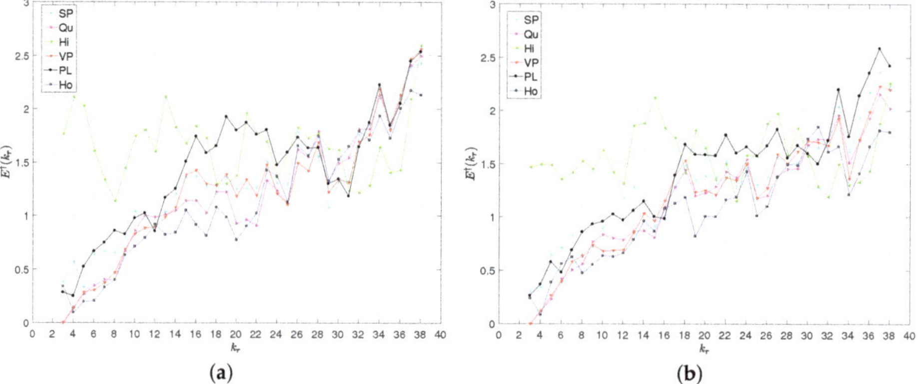

Figure 2. The error $E^{\dagger}(k_r)$, $k_r = 3, \ldots, 38$, for the models (1a)–(1f), during the season 2018–2019: (**a**) 'La liga'; (**b**) 'Premiership'.

Table 1. The mean and standard deviation, μ and σ, of the $E^{\dagger}(k_r)$ series for the four leagues and the models (1a)–(1f), during the season 2018–2019.

		SP	Qu	Hi	VP	PL	Ho
'La Liga'	μ	1.3182	1.1832	1.5852	1.2051	1.4118	1.0782
	σ	0.4986	0.5441	0.2592	0.5607	0.5746	0.4736
'Premiership'	μ	1.2354	1.1082	1.5061	1.1699	1.3368	1.0394
	σ	0.5233	0.5529	0.2240	0.5984	0.5496	0.5112
'Serie A'	μ	1.1749	1.0574	1.5783	1.1596	1.3800	1.0045
	σ	0.4426	0.4865	0.2966	0.5977	0.5963	0.4251
'Ligue 1'	μ	1.3035	1.2053	1.6562	1.2313	1.4357	1.1333
	σ	0.5155	0.6144	0.3093	0.6080	0.5627	0.5813

Figure 3 depicts the parameters $\{a_1(k_r), b_1(k_r)\}$ of the PL and $\{a_1(k_r), b_1(k_r), c_1(k_r)\}$ for the Ho model for FC Barcelona and Manchester City, respectively. The point labels represent the value of k_r. We verify that for the FC Barcelona we have two distinct periods in the locus. The first corresponds to

$3 \leq k_r \leq 8$, where the parameters evolve influenced by a set of consecutive bad results between rounds five and eight. The second period corresponds to $8 \leq k_r \leq 38$, where the path changes direction driven by a consistent and positive team behavior towards the final victory at $k_r = 38$. For the Manchester City the variation of the parameters is more complex. Initially, we observe a route for the period $3 \leq k_r \leq 7$. Then, the locus has a slight change, due to a draw achieved by the team at round 8, but recovers fast its initial trend during for $9 \leq k_r \leq 15$. Again the locus changes driven by the set of team negative results in rounds 16–19 and 24. From $k_r = 25$ onward, the parameters evolve positively influenced by the consecutive team victories until the end of the season at $k_r = 38$. For other teams we can draw similar conclusions, meaning that there exists a clear relationship between the models' parameters and the teams' performance along the season. Moreover, we verify that, in general, abrupt changes in the loci route correspond to inconsistent results at early rounds, that is, small values of k_r. For larger values of k_r, eventual inconsistencies on the teams' behavior do not translate in significant modifications of the parameter patterns, since the fitting becomes less sensitive to the number of fitting points.

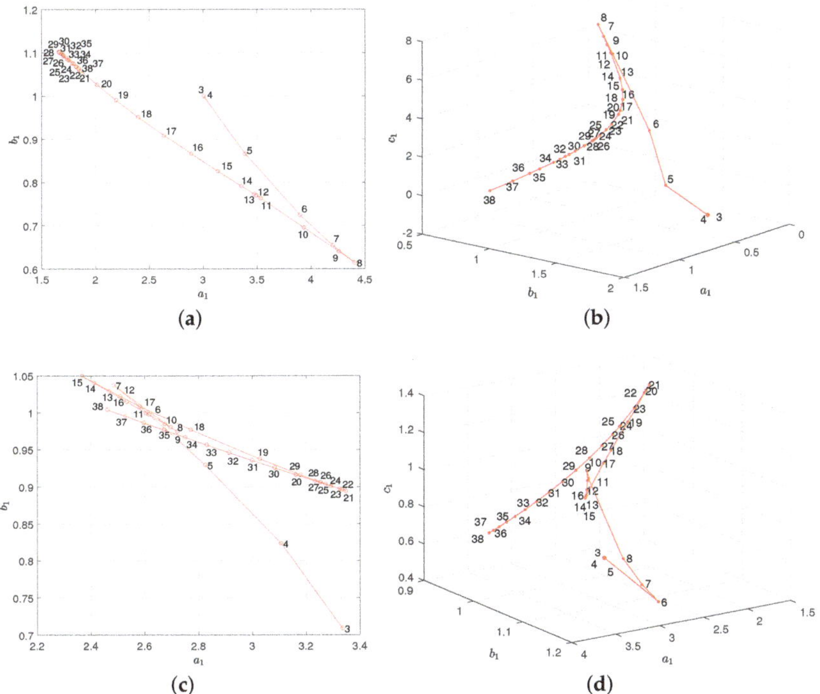

Figure 3. Locus of the the models parameters for the 2018–2019 champions of 'La Liga' and 'Premiership': (**a**) PL model for FC Barcelona; (**b**) Ho model for FC Barcelona; (**c**) PL model for Manchester City; (**d**) Ho model for Manchester City. The point labels denote $k_r = 3, \ldots, 38$.

3. Entropy of the Spatio-Temporal Patterns of the Models' Parameters

In this section we characterize the soccer leagues by means of entropy. The entropy is a measure of regularity that has been successfully adopted in the study of complex systems [15,32].

3.1. The Entropy of the PL Model

By approximating the output signals $x_i(k)$ through PL functions (1e) we are modeling the complex system as a fractional integrator [33,34] of order $b_i \in \mathbb{R}^+$ for a constant, step-like, input signal.

If a team obtains {victory, draw, defeat} in all matches, then $x_i(k)$ is a straight line with $a_i = \{3, 1, 0\}$ and $b_i = 1$. However, real-world teams have {victories, draws, defeats} and, thus, yield a fractal like response that follows a PL behavior. Therefore, fractional/unit values of b_i reflect variable/constant time evolution, while values of a_i close to $\{3, 1, 0\}$ correspond to {victory, draw, defeat} results [10].

For each league we now compute the PL parameters $\{a_i, b_i\}$ that fit the teams' positions $x_i(k)$, $i = 1, \ldots, 20$, $0 \leq k \leq k_r$, from the beginning of the season up to the round $k_r = 3, \ldots, 38$. Therefore, for every k_r we have an array of $20 \times (k_r - 2)$ points in a two-dimensional space. We then determine the bi-dimensional histograms by binning the data of each array into $M \times M = 100 \times 100$ bins $\{\alpha_j, \beta_k\}$, $j, k = 1, \ldots, M$. Finally, we calculate the Shannon entropy [35,36]:

$$S(k_r) = -\sum_{j=1}^{M}\sum_{k=1}^{M} P\left(\alpha_j, \beta_k\right) \log P\left(\alpha_j, \beta_k\right), \tag{3}$$

where the probabilities $P\left(\alpha_j, \beta_k\right)$ are approximated by the data relative frequencies.

For example, Figure 4 depicts the histograms of the PL parameters from the beginning, $k_r = 3$, up to the end of the 2018–2019 season, $k_r = 38$, for 'La Liga', 'Premiership', 'Serie A' and 'Ligue 1'. We verify that the parameters $\{a_i, b_i\}$ exhibit less dispersion for the pair $\mathcal{P}_1 = \{$'La Liga', 'Ligue 1'$\}$ than for the pair $\mathcal{P}_2 = \{$'Premiership', 'Serie A'$\}$.

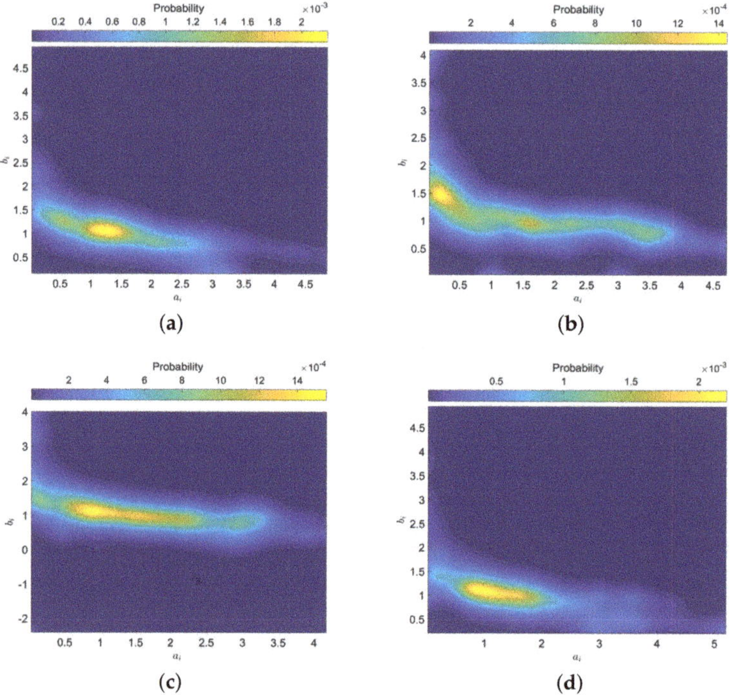

Figure 4. Histograms of the PL parameters $\{a_i, b_i\}$ from the beginning, $k_r = 3$, up to the end of the 2018–2019 season, $k_r = 38$, and the leagues: (**a**) 'La Liga'; (**b**) 'Premiership'; (**c**) 'Serie A'; (**d**) 'Ligue 1'.

Figure 5 illustrates the evolution on the entropy, $S(k_r)$ versus $k_r = 3, \ldots, 38$, for the PL parameters and 'La Liga', 'Premiership', 'Serie A' and 'Ligue 1'. Again, we verify that the pairs $\mathcal{P}_1$ and $\mathcal{P}_2$ reveal similar behavior. For the pair $\mathcal{P}_1$, the entropy increases faster with k_r that for the pair $\mathcal{P}_2$. This means that, in the 2018–2019 season, the Spanish and French teams started more irregular than the English and Italian ones. Nevertheless, since for all leagues, $S(k_r)$ converges to a similar settling value, we conclude that by the end of the season the {victory, draw, defeat} global pattern exhibited by teams in different leagues is identical.

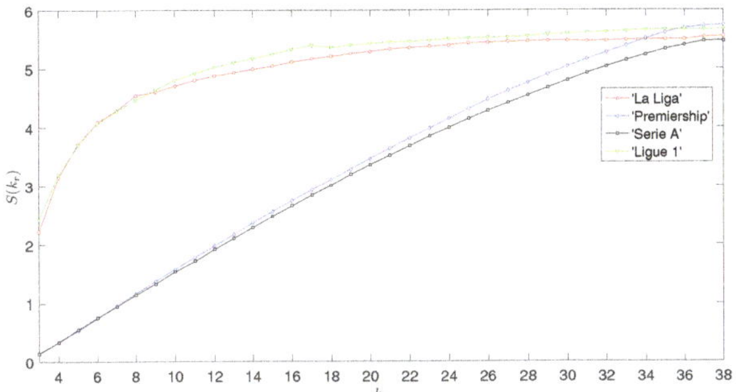

Figure 5. Evolution of the entropy $S(k_r)$ versus $k_r = 3, \ldots, 38$, for the PL parameters and the 'La Liga', 'Premiership', 'Serie A' and 'Ligue 1', during the season 2018–2019.

The symmetry between the pairs $\mathcal{P}_1$ and $\mathcal{P}_2$, both in the histograms of Figure 4 and the entropy evolution represented in Figure 5 opens, however, new questions. Is such duality of patterns just a casual result, or does it point towards some underlying effects ruling the dynamics of these complex systems? Additionally, can other patterns occur and what is their meaning? A future study addressing a larger number of leagues seems necessary in order to give a response to these type of questions.

3.2. The Entropy of the Ho Model

The Ho model (1f) combines both an exponential and a PL function. These functions characterize well integer- and fractional-order systems, respectively [37].

Similarly to the previous subsection, we compute the Ho parameters $\{a_i, b_i, c_i\}$ that fit the teams' positions $x_i(k), i = 1, \ldots, 20, 0 \leq k \leq k_r$, from the beginning of the season up to the round $k_r = 3, \ldots, 38$. For every k_r we now obtain an array of $20 \times (k_r - 2)$ points in a three-dimensional space. Therefore, we determine three-dimensional histograms by binning the data of each array into $M \times M \times M = 100 \times 100 \times 100$ bins $\{\alpha_j, \beta_k, \gamma_l\}, j, k, l = 1, \ldots, M$. Finally, we calculate the Shannon entropy [35,36]:

$$S(k_r) = -\sum_{j=1}^{M} \sum_{k=1}^{M} \sum_{l=1}^{M} P\left(\alpha_j, \beta_k, \gamma_l\right) \log P\left(\alpha_j, \beta_k, \gamma_l\right), \tag{4}$$

where the probabilities $P\left(\alpha_j, \beta_k, \gamma_l\right)$ are approximated by the data relative frequencies.

Figure 6 depicts the two-dimensional projections of the histogram of the Ho parameters, from the beginning, $k_r = 3$, up to the end of the 2018–2019 season, $k_r = 38$, for 'La Liga', 'Premiership', 'Serie A' and 'Ligue 1'. Therefore, the charts represent the combination of the pairs of parameters $\{a_i, b_i\}$, $\{a_i, c_i\}$ and $\{b_i, c_i\}$. As for the PL model, we verify that, in general, the parameters $\{a_i, b_i, c_i\}$ exhibit less dispersion for the leagues $\mathcal{P}_1 = \{$'La Liga', 'Ligue 1'$\}$ than for the leagues $\mathcal{P}_2 = \{$'Premiership', 'Serie A'$\}$.

Figure 7 depicts the entropy, $S(k_r)$ versus $k_r = 3, \ldots, 38$, for the Ho model parameters, and 'La Liga', 'Premiership', 'Serie A' and 'Ligue 1'. We verify that $S(k_r)$ has now an identical behavior for all leagues, but we can still distinguish a slight difference between the pairs $\mathcal{P}_1$ and $\mathcal{P}_2$.

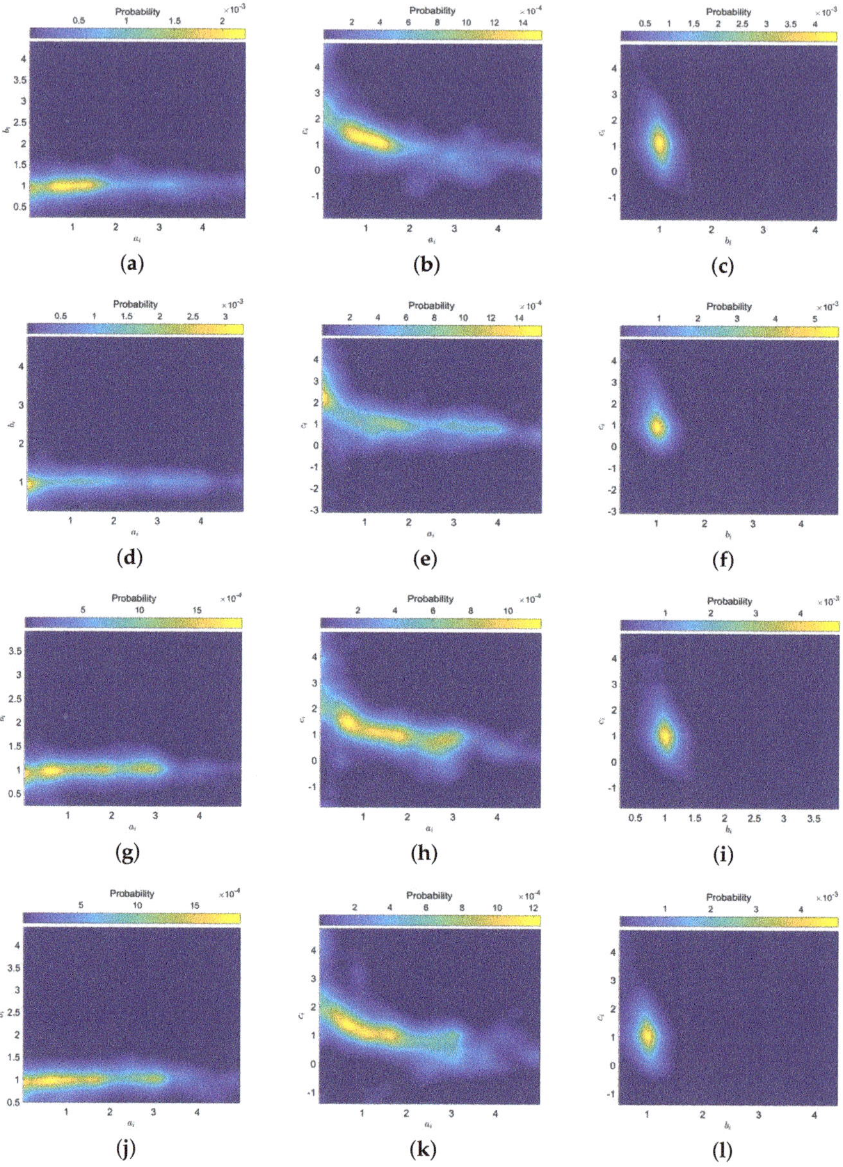

Figure 6. The two-dimensional projections of the histograms of the Ho parameters, from the beginning, $k_r = 3$, up to the end of the 2018–2019 season, and the leagues: (**a–c**) 'La Liga'; (**d–f**) 'Premiership'; (**g–i**) 'Serie A'; (**j–l**) 'Ligue 1'. The charts represent the combination of the pairs of parameters $\{a_i, b_i\}$, $\{a_i, c_i\}$ and $\{b_i, c_i\}$.

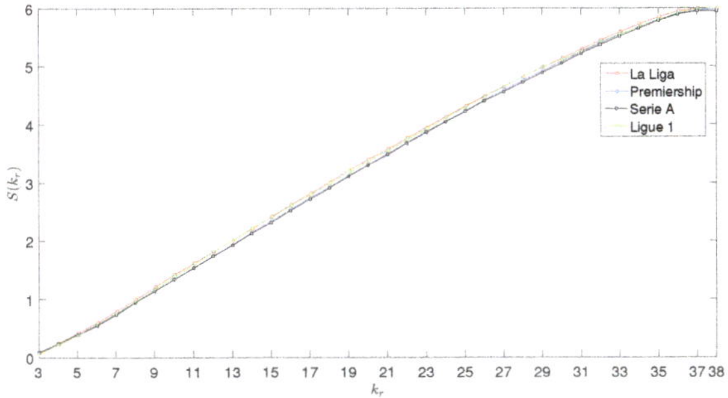

Figure 7. Evolution of the entropy $S(k_r)$ versus $k_r = 3, \ldots, 38$, for the Ho model parameters and the leagues 'La Liga', 'Premiership', 'Serie A' and 'Ligue 1', during the season 2018–2019.

4. Predicting the Teams' Results

The models (1a)–(1f), introduced in section 2, are now tested in the prediction of the teams' results. In a first phase, we fit the models to the series $x_i(k)$, $i = 1, \ldots, N$, $0 \leq k \leq k_r$, and we calculate the approximation $\hat{x}_i(k)$ for each $k_r = 3, \ldots, 37$. In a second phase, we extrapolate the values of $\hat{x}_i(k+1)$, that is, the teams positions for every round from four to 38. Finally, in a third phase, we assess the accuracy of the values $\hat{x}_i(k+1)$ by means of the prediction errors E and $E^{\dagger}$, where $k_r = 4, \ldots, 38$.

Figure 8 illustrates the error $E_1(k_r)$, $k_r = 4, \ldots, 38$, obtained with the models (1a)–(1f) for the champions of the 'La Liga', 'Premiership', 'Serie A' and 'Ligue 1' in season 2018–2019, namely for the FC Barcelona, Manchester City, Juventus and Paris Saint-Germain. Figure 9 depicts $E^{\dagger}(k_r)$, $k_r = 4, \ldots, 38$, for 'La liga', 'Premiership', 'Serie A' and 'Ligue 1', obtained with the six models (1a)–(1f). We verify again that, with the exception of the Hi model (1c), all other models predict well the teams' results.

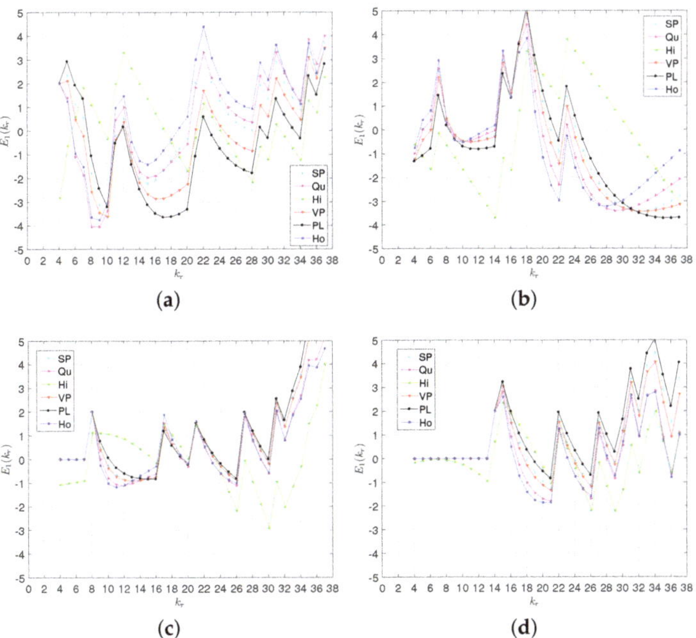

Figure 8. The error $E_1(k_r)$, $k_r = 4, \ldots, 38$, for the models (1a)–(1f), during the season 2018–2019: (**a**) FC Barcelona; (**b**) Manchester City; (**c**) Juventus; (**d**) Paris Saint-Germain.

For a numerical comparison of the models accuracy, we calculate the mean and the standard deviation, μ and σ, of the $E^{\dagger}(k_r)$ series and list their values in Table 2. We verify again that the Ho (1f) model leads to the best predictions. We can see that the errors are larger for the leagues in the set $\mathcal{P}_1$ than for the ones in the set $\mathcal{P}_2$, meaning that the prediction is more difficult for the leagues with higher values of entropy.

Table 2. The mean and standard deviation, μ and σ, of the $E^{\dagger}(k_r)$ for the six models (1a)–(1f) and season 2018–2019.

		SP	Qu	Hi	VP	PL	Ho
'La Liga'	μ	2.0137	2.1177	1.9397	2.0604	2.1631	1.9172
	σ	0.3807	0.4339	0.3076	0.3945	0.3684	0.4173
'Premiership'	μ	1.9765	1.9543	1.9936	2.0860	2.0926	1.9036
	σ	0.5519	0.4019	0.2198	0.6336	0.5226	0.6672
'Serie A'	μ	1.8093	1.8717	1.8765	1.9462	2.0468	1.7322
	σ	0.2930	0.2764	0.3042	0.4216	0.4139	0.5888
'Ligue 1'	μ	2.0407	2.1022	1.9912	2.1400	2.1935	1.9337
	σ	0.3226	0.2835	0.2657	0.4683	0.4653	0.5536

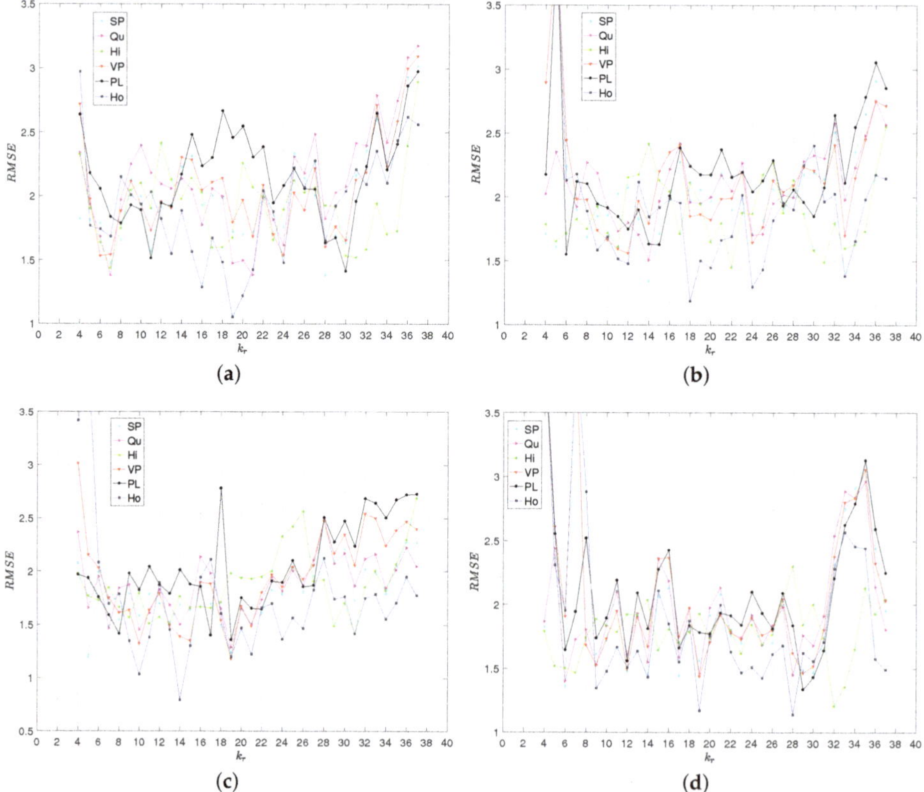

Figure 9. The error $E^{\dagger}(k_r)$, $k_r = 4, \ldots, 38$, for the models (1a)–(1f), during the season 2018–2019: (**a**) 'La liga'; (**b**) 'Premiership'; (**c**) 'Serie A'; (**d**) 'Ligue 1'.

5. Conclusions

We proposed a fractional systems' perspective for analyzing soccer teams competing within a league season. Firstly, we adopted six fitting models to describe the teams' positions along one season and interpreted the loci of the models' parameters as a signature of the system dynamics. Secondly, we studied the entropy of the models parameters' spatio-temporal patterns for comparing different leagues. Both approaches represent valid tools to describe the complex behavior of such challenging systems. The computational modeling unraveled patterns embedded in the data suggesting some common underlying dynamical effects in different leagues. The prediction quality of the two models, both in the perspective of each individual team and the league, along the season, was also analyzed. Nonetheless, several new questions emerged in the sequence of the statistical and entropic analysis. Is the apparent duality between the pairs $\mathcal{P}_1$ and $\mathcal{P}_2$ just some coincidence or do they reflect some kind of additional effects besides the standard rules of the game? The investigation of these and other questions needs the future algorithmic treatment of more data involving more seasons and leagues.

Author Contributions: A.M.L. and J.A.T.M. conceived, designed and performed the experiments, analyzed the data and wrote the paper. These authors contributed equally to this work. All authors have read and agreed to the published version of the manuscript.

Funding: This research received no external funding.

Conflicts of Interest: The authors declare no conflict of interest.

References

1. Carling, C.; Williams, A.M.; Reilly, T. *Handbook of Soccer Match Analysis: A Systematic Approach to Improving Performance*; Routledge: Abingdon, UK, 2007.
2. Giulianotti, R. Football. In *The Wiley-Blackwell Encyclopedia of Globalization*; Wiley: Hoboken, NJ, USA, 2012.
3. Brocas, I.; Carrillo, J.D. Do the "three-point victory" and "golden goal" rules make soccer more exciting? *J. Sport. Econ.* **2004**, *5*, 169–185. [CrossRef]
4. Miller, K.S.; Ross, B. *An Introduction to the Fractional Calculus and Fractional Differential Equations*; Wiley: Hoboken, NJ, USA, 1993.
5. Tenreiro Machado, J.A.; Kiryakova, V. Recent history of the fractional calculus: Data and statistics. In *Handbook of Fractional Calculus with Applications: Basic Theory*; Kochubei, A.; Luchko, Y., Eds.; De Gruyter: Berlin, Germany, 2019; Volume 1, pp. 1–21.
6. Machado, J.T.; Pinto, C.M.; Lopes, A.M. A review on the characterization of signals and systems by power law distributions. *Signal Process.* **2015**, *107*, 246–253. [CrossRef]
7. Parsa, B.; Dabiri, A.; Machado, J.A.T. Application of Variable order Fractional Calculus in Solid Mechanics. In *Handbook of Fractional Calculus with Applications: Applications in Engineering, Life and Social Sciences, Part A*; Baleanu, D.; Lopes, A.M., Eds.; De Gruyter: Berlin, Germany, 2019; Volume 7, pp. 207–224.
8. Machado, J.T.; Lopes, A.M. Fractional Electromagnetics. In *Handbook of Fractional Calculus with Applications: Applications in Physics, Part B*; Tarasov, V.E., Ed.; De Gruyter: Berlin, Germany, 2019; Volume 5, pp. 1–23.
9. Lopes, A.M.; Machado, J. Fractional order models of leaves. *J. Vib. Control.* **2014**, *20*, 998–1008. [CrossRef]
10. Machado, J.T.; Lopes, A.M. On the mathematical modeling of soccer dynamics. *Commun. Nonlinear Sci. Numer. Simul.* **2017**, *53*, 142–153. [CrossRef]
11. Machado, J.; Mata, M.E.; Lopes, A.M. Fractional state space analysis of economic systems. *Entropy* **2015**, *17*, 5402–5421. [CrossRef]
12. Valério, D.; Ortigueira, M.; Machado, J.T.; Lopes, A.M. Continuous-time fractional linear systems: Steady-state behaviour. In *Handbook of Fractional Calculus with Applications: Applications in Engineering, Life and Social Sciences, Part A*; Petráš, I., Ed.; De Gruyter: Berlin, Germany, 2019; Volume 6, pp. 149–174.
13. Couceiro, M.S.; Clemente, F.M.; Martins, F.M.; Machado, J.A.T. Dynamical stability and predictability of football players: The study of one match. *Entropy* **2014**, *16*, 645–674. [CrossRef]
14. Silva, P.; Duarte, R.; Esteves, P.; Travassos, B.; Vilar, L. Application of entropy measures to analysis of performance in team sports. *Int. J. Perform. Anal. Sport* **2016**, *16*, 753–768. [CrossRef]

15. Machado, J.T.; Lopes, A.M. Multidimensional scaling analysis of soccer dynamics. *Appl. Math. Model.* **2017**, *45*, 642–652. [CrossRef]
16. Neuman, Y.; Israeli, N.; Vilenchik, D.; Cohen, Y. The adaptive behavior of a soccer team: An entropy-based analysis. *Entropy* **2018**, *20*, 758. [CrossRef]
17. Lopes, A.M.; Tenreiro Machado, J. Entropy Analysis of Soccer Dynamics. *Entropy* **2019**, *21*, 187. [CrossRef]
18. Berrar, D.; Lopes, P.; Davis, J.; Dubitzky, W. Guest editorial: special issue on machine learning for soccer. *Mach. Learn.* **2019**, *108*, 1–7. [CrossRef]
19. Karlis, D.; Ntzoufras, I. Analysis of sports data by using bivariate Poisson models. *J. R. Stat. Soc. Ser. D (The Stat.)* **2003**, *52*, 381–393. [CrossRef]
20. Baio, G.; Blangiardo, M. Bayesian hierarchical model for the prediction of football results. *J. Appl. Stat.* **2010**, *37*, 253–264. [CrossRef]
21. Hvattum, L.M.; Arntzen, H. Using ELO ratings for match result prediction in association football. *Int. J. Forecast.* **2010**, *26*, 460–470. [CrossRef]
22. Berrar, D.; Lopes, P.; Dubitzky, W. Incorporating domain knowledge in machine learning for soccer outcome prediction. *Mach. Learn.* **2019**, *108*, 97–126. [CrossRef]
23. Murphy, K.P. *Machine Learning: A Probabilistic Perspective*; MIT Press: Cambridge, MA, USA, 2012.
24. Van Haaren, J.; Van den Broeck, G. Relational learning for football-related predictions. In *Latest Advances in Inductive Logic Programming*; World Scientific: Singapore, 2015; pp. 237–244.
25. Hubáček, O.; Šourek, G.; Železný, F. Learning to predict soccer results from relational data with gradient boosted trees. *Mach. Learn.* **2019**, *108*, 29–47. [CrossRef]
26. Tsokos, A.; Narayanan, S.; Kosmidis, I.; Baio, G.; Cucuringu, M.; Whitaker, G.; Király, F. Modeling outcomes of soccer matches. *Mach. Learn.* **2019**, *108*, 77–95. [CrossRef]
27. Edla, D.R.; Lingras, P.; Venkatanareshbabu, K. *Advances in Machine Learning and Data Science: Recent Achievements and Research Directives*; Springer: Berlin, Germany, 2018.
28. Abu-Mostafa, Y.S.; Magdon-Ismail, M.; Lin, H.T. *Learning from Data*; AMLBook: New York, NY, USA, 2012.
29. Lawson, C.L.; Hanson, R.J. *Solving Least Squares Problems*; SIAM: Philadelphia, PA, USA, 1974; Volume 161.
30. Draper, N.R.; Smith, H.; Pownell, E. *Applied Regression Analysis*; Wiley: New York, NY, USA, 1966; Volume 3.
31. Lopes, A.; Tenreiro Machado, J.; Galhano, A. Empirical laws and foreseeing the future of technological progress. *Entropy* **2016**, *18*, 217. [CrossRef]
32. Lopes, A.; Tenreiro Machado, J. Complexity Analysis of Global Temperature Time Series. *Entropy* **2018**, *20*, 437. [CrossRef]
33. Machado, J.T.; Lopes, A.M. The persistence of memory. *Nonlinear Dyn.* **2015**, *79*, 63–82. [CrossRef]
34. Machado, J.; Lopes, A.M. Analysis of natural and artificial phenomena using signal processing and fractional calculus. *Fract. Calc. Appl. Anal.* **2015**, *18*, 459–478. [CrossRef]
35. Shannon, C. A Mathematical Theory of Communication. *Bell Syst. Tech. J.* **1948**, *27*, 379–423. [CrossRef]
36. Jaynes, E. Information Theory and Statistical Mechanics. *Phys. Rev.* **1957**, *106*, 620–630. [CrossRef]
37. Pinto, C.M.; Lopes, A.M.; Machado, J.T. A review of power laws in real life phenomena. *Commun. Nonlinear Sci. Numer. Simul.* **2012**, *17*, 3558–3578. [CrossRef]

Article

Puu System of Fractional Order and Its Chaos Suppression

Marius-F. Danca [1,2]

[1] Department of Mathematics and Computer Science, Avram Iancu University, 400380 Cluj-Napoca, Romania; danca@rist.ro

[2] Romanian Institute of Science and Technology, 400487 Cluj-Napoca, Romania

Received: 5 February 2020; Accepted: 14 February 2020; Published: 27 February 2020

Abstract: In this paper, the fractional-order variant of Puu's system is introduced, and, comparatively with its integer-order counterpart, some of its characteristics are presented. Next, an impulsive chaos control algorithm is applied to suppress the chaos. Because fractional-order continuous-time or discrete-time systems have not had non-constant periodic solutions, chaos suppression is considered under some numerical assumptions.

Keywords: Puu system of fractional order; Caputo-like delta difference; impulsive control; chaos suppression

---

1. Introduction

The impulsive control concept has a long history that is based on the mathematical foundation of impulsive differential equations. Most examples of impulsive phenomena can be found in mechanical systems with impacts where a sudden change in their states appears, but also in population dynamics, biotechnology processes, chemistry, engineering, medicine, spacecraft optimal control, and so on. Impulsive systems can be studied via the mathematical tool based on impulsive differential or discrete equations. In the last few decades, differential equations with impulses can be found in e.g., nanoelectronic devices, chaotic spread-spectrum communication systems, or electrical engineering applications, and so on (see, e.g., [1,2]). On the other side, there are classes of systems like biological systems, economical systems where discrete time models seem to be more realistic than the continuous ones.

There exist several kinds of impulses [3]—for example, systems where the impulses are applied at fixed time-moments (see, e.g., one of the first references [4]) and also systems where the impulses are applied at variable times [5].

Due to the memory and hereditary properties of the fractional derivatives (see [6,7]), the discrete-time or continuous-time systems of fractional order (FO) are more suitable than integer-order systems.

While the definition of fractional derivative for continuous-time real functions has been formulated in the late 19th Century by Liouville, Grunwald, Letnikov, and Riemann, the first definition of a fractional difference operator was made by Diaz and Olser in 1974 [8]. Nowadays, differential or difference equations of FO represent useful models in viscoelasticity, mechatronics, seismology, aerodynamics, electrical circuits, biophysics, biology, blood flow phenomena, chemistry, control theory, etc (see, e.g., [9]).

In this paper, the fractional order (FO) variant of the Puu's system is numerically analyzed and compared with its integer order (IO) counterpart. Using the results on non-existence of exact periodic solutions of FO discrete systems, one consider the periodicity as computationally approached. Moreover, using an impulsive algorithm, the chaotic motion can be suppressed.

2. Puu's Fractional Order System

In 1939, Paul Samuelson [10] introduced the standard principle of acceleration to consumption used in one of the first formal mathematical models for business cycles, while Sir John Hicks (1950) later improved the Samuelson model [11], introducing the "floor" and "roof" to depreciation levels.

Starting from the standard Samuelson-Hicks model, in 1989, Puu and Sushko [12] developed the discrete dynamical income system of integer order (IO) with cubic nonlinearity of integer order, modeled by the following cubic initial value problem (IVP)

$$x(n+1) = ax(n) - (a+1)x^3(n),\ x(0) = x_0, \tag{1}$$

with $a > 0$ a real parameter.

It is usual to have multiple attractors for a dynamical system. Each of them can be considered as the attractor for a given initial condition within its own attraction basin. In the bifurcation diagram vs. $a \in [0,3]$, presented in Figure 1a, the Puu system of IO (1) reveals two symmetric regions representing regular and chaotic attractors generated by two symmetric (negative and positive) initial conditions within two attraction basins (Due to its symmetry, the coexistence of symmetric attractors is a feature of the Puu system. Like for the cubic logistic map $x \to kx(1-x^2)$, the oddness of the map (1) induces an equivariance that is a $\mathbb{Z}_2$-symmetry [13,14].). Therefore, attractors corresponding to $a \in [0,3]$ in the upper (bottom) region present non-symmetric new generated points. The successive bifurcations of these points lead to chaos via the standard period-doubling cascade. In addition, note the sudden crisis—at $a = 2.6$, the previous two non-symmetric chaotic attractors (red and blue, respectively) collide and give birth to a symmetric chaotic attractor that covers both positive and negative values of x. Details on similar dynamics but related to the cubic logistic map can be found in [15].

Let us consider the FO Puu discrete system. Following the approach in [16] (see also [17], where the synchronization of the FO Puu is considered), the Caputo-like Puu system of FO can be modeled by the following general IVP of FO:

$$\Delta_*^q x(k) = f(x(k+q-1)),\ k \in \mathbb{N}_{1-q},\ x(0) = x_0, \tag{2}$$

where Δ_*^q is the Caputo-like delta difference of order $0 < q \leq 1$, and $N_{1-q} = \{1-q, 2-q, ...,\}$ represents the isolated time scale. With the cubic right-hand side $f(x) = x - (a+1)x^3$, the IVP (2) reads

$$\Delta_*^q x(k) = ax(k+q-1) - (a+1)x^3(k+q-1),\ k \in \mathbb{N}_{1-q},\ x(0) = x_0, \tag{3}$$

Theorem. *[16,18] The IVP (3) admits the following discrete integral*

$$x(n) = x(0) + \frac{1}{\Gamma(q)} \sum_{j=1-q}^{n-q} \frac{\Gamma(n-j)}{\Gamma(n-j-q)} [ax(j+q-1) - (a+1)x^3(j+q-1)]. \tag{4}$$

In (4), Γ represents the Gamma function.

Numerically implemented, via the substitution $i = j + q$, the integral (4) can be expressed as follows [16]:

$$x(n) = x(0) + \frac{1}{\Gamma(q)} \sum_{i=1}^{n} \frac{\Gamma(n-i+q)}{\Gamma(n-i+1)} [ax(i-1) - (a+1)x^3(i-1)],\ n = 1, 2, ... \tag{5}$$

The local Lyapunov exponent λ, can be approximated numerically at the first n iterations with the following relation [19]:

$$\lambda = \frac{1}{n} \ln(|\tau(n-1)|), \tag{6}$$

where the tangent map $\tau(n)$, obtained by the linearization of (5) along the orbit $x(n)$, is

$$\tau(n) = \tau(0) + \frac{1}{\Gamma(q)} \sum_{j=1}^{n} \frac{\Gamma(n-i+q)}{\Gamma(n-i+1)} [a\tau(j-1) - 3(a+1)x^2(j-1)\tau(j-1)], \ \tau(0) = 1.$$

Like the solution x, the tangent map τ at the moment n also presents the so-called discrete memory effect (or time history) i.e., the numerical determined value at the moment n depends on all previous values. Because in relations (5) and (6), the term $R := \sum_{j=1}^{n} \frac{\Gamma(n-i+q)}{\Gamma(n-i+1)}$ presents divergency problems, to deal numerically with R for large values of n, one can use the relation $\frac{\Gamma(n-i+q)}{\Gamma(n-i+1)} = e^{\ln(\Gamma(n-i+q)) - \ln(\Gamma(n-i+1))}$. In this way, the numerical experiments can be extended for n up to a couple thousand.

The bifurcation diagrams of the Puu system of FO vs. q, generated within $x_0 = \pm 0.085$ and $x_0 = \pm 0.8$, for $a = 1.27$ and $a = 1.14$ (Figure 1b,c, respectively), reveal symmetrical transitions to chaos and, also, contrary to its IO part, attractors' coexistence.

Hereafter, only the two merged "positive" regions (blue and cyan) located most entirely in the positive axis are considered. The other two "negative" symmetric regions (red and light blue) can be generated with $x_0 = -0.085$ and $x_0 = -0.8$, respectively. As can be seen, both "positive" and "negative" attractors corresponding to $q \in (0, 1]$ maintain the odd symmetry existing in the integer order system and the bifurcations in the positive (negative) axis are non-symmetric.

Remark 1. *As for the case of fractional differential equations, contrary to integer order difference equations, fractional difference equations do not admit exact periodic but only asymptotically periodic solutions (see, e.g., [20–22], respectively). Therefore, notions like "stable cycle", "chaos control", and even "chaos" (as consisting in infinitely many unstable* periodic *motions), in continuous-time or discrete-time systems of FO seem to not be adequate. Up to a considered enough tiny computational error, the apparent periodic orbits are called here "computationally periodic orbits" (CPOs). Therefore, in this paper, the chaos suppression means to obtain stable CPOs.*

To study some orbit $x(n)$, the relations (5) and (6) together with time series, histograms, and the '0–1' test [23] will be used. Because we are interested in chaos suppression, the value of the parameter a is chosen so that the system evolves chaotic for all initial conditions. Thus, for $a = 1.3$, the system behaves chaotically for all initial conditions.

One of the tools generated by the '0–1' test [23], which can be easily numerically implemented starting from a time series related to the considered dynamical continuous or discrete system, is the asymptotic growth rate, K, which gives important information to distinguish chaotic behavior from regular behavior. Note that the '0-1' test does not need the system equations, but only a time series of the system. If $K \approx 1$, then the underlying dynamics are chaotic, while, if $K \approx 0$, the system behaves along some stable CPO. It is commonly assumed that about 2000 elements in the considered time series, after transients are discarded, are enough. In this paper, after the first 1000 iterations neglected, 1500 elements give enough accurate results.

While the integer order system develops, as usual, stronger chaos once the bifurcation parameter a increases (Figure 1a), the FO system presents chaos extinction with the increase of a (Figure 1b,c).

Hereafter, the case $q = 0.5$ will be considered.

Another interesting property of the FO Puu system, contrary to its integer order counterpart, is the coexistence of stable CPOs with chaotic attractors. For example, in Figure 2, two zoomed areas are presented (Figure 2b,c) from the bifurcation diagram for $a \in [0, 2]$ in Figure 2a. Both zoomed areas reveal, for some parameter values of a, the coexistence of a chaotic attractor with a stable CPO (dotted lines at $a = 1.27$ and $a = 1.14$).

Consider $a = 1.27$ within a stable CP window of period-5 for $a \in [1.22, 1.32]$ (red plot) coexisting with its chaotic counterpart (blue plot) in Figure 2b. The coexistence of the underlying stable CPO and the chaotic attractor is studied in Figure 3. Figure 3a,b present superimposed, the time series entirely and the last 100 iterations, respectively; in Figure 3c,d, superimposed, the Lyapunov exponent λ and the asymptotic growth rate, K, are represented, while in Figure 3e,f plot the histograms of the coexisting attractors. The chaotic attractor generated from $x_0 = 0.8$ (blue plot in Figure 3a,b is characterized by the evolution of the time series, the K value, which is close to 1, and the positiveness of the superimposed Lyapunov exponent λ (Figure 3d) and the histogram in Figure 3f. The stable CPO revealed after about 900 neglected iterations from the considered 2500 iterations, generated from $x_0 = 0.085$ (red plot in Figure 3a,b), is characterized by the evolution of the time series, the K value, and the superimposed λ, in which both are close to 0 (Figure 3c), and the discrete five peaks bars in the histogram (Figure 3e) indicating the branches of the cycle, numbered 1–5 in the zoomed time series in Figure 3b. Note that the zero value of λ seems to indicate a kind of Neimark–Sacker bifurcation, generating the stable CPO of period-5 which coexists with the chaotic attractor. In addition, as can be seen in Figure 3c,d, this kind of bifurcation seems to be not singular, but this subject does not represent the purpose of this paper.

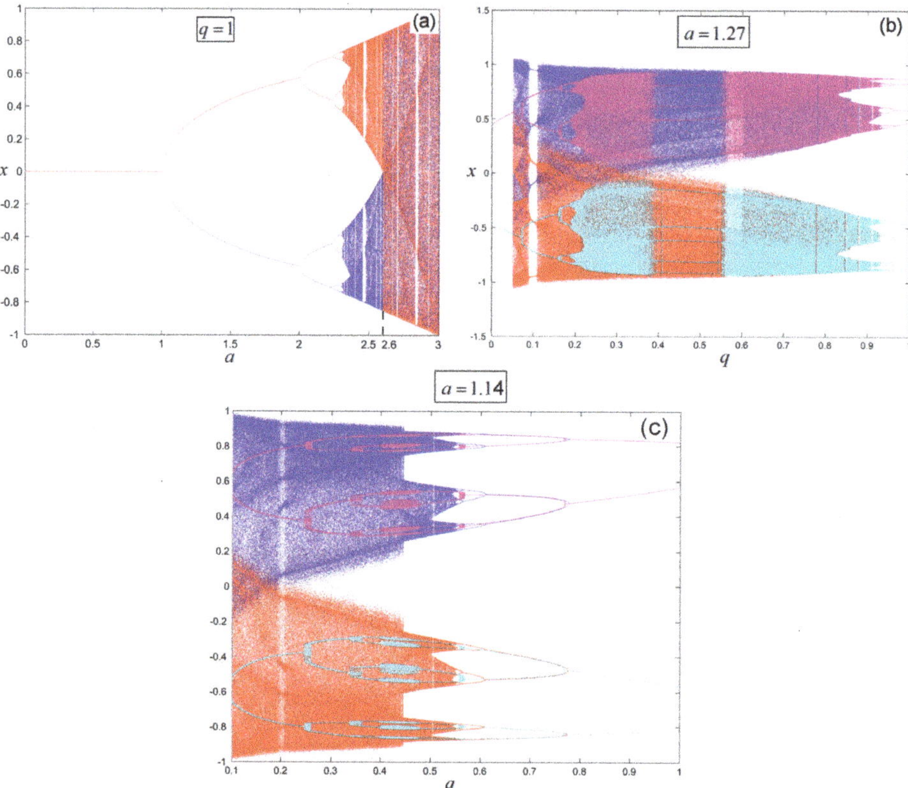

Figure 1. Bifurcation diagrams of the Puu system of IO and FO. (**a**) bifurcation of the Puu system of IO vs. parameter a; (**b**) bifurcation of the Puu system of FO vs. q for $a = 1.27$; (**c**) bifurcation of the Puu system of FO for $a = 1.14$.

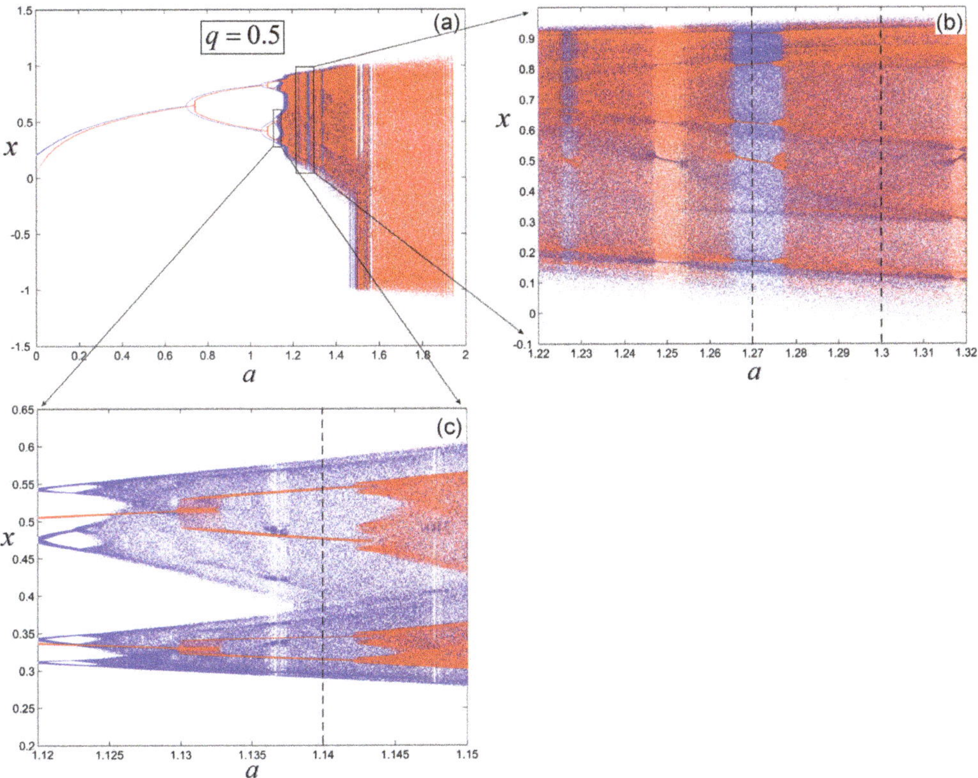

Figure 2. Bifurcations of the Puu system of FO. (**a**) bifurcation vs. a for $q = 0.5$; (**b**) zoomed region for $a \in [1.22, 1.32]$; (**c**) zoomed region for $a \in [1.12, 1.15]$.

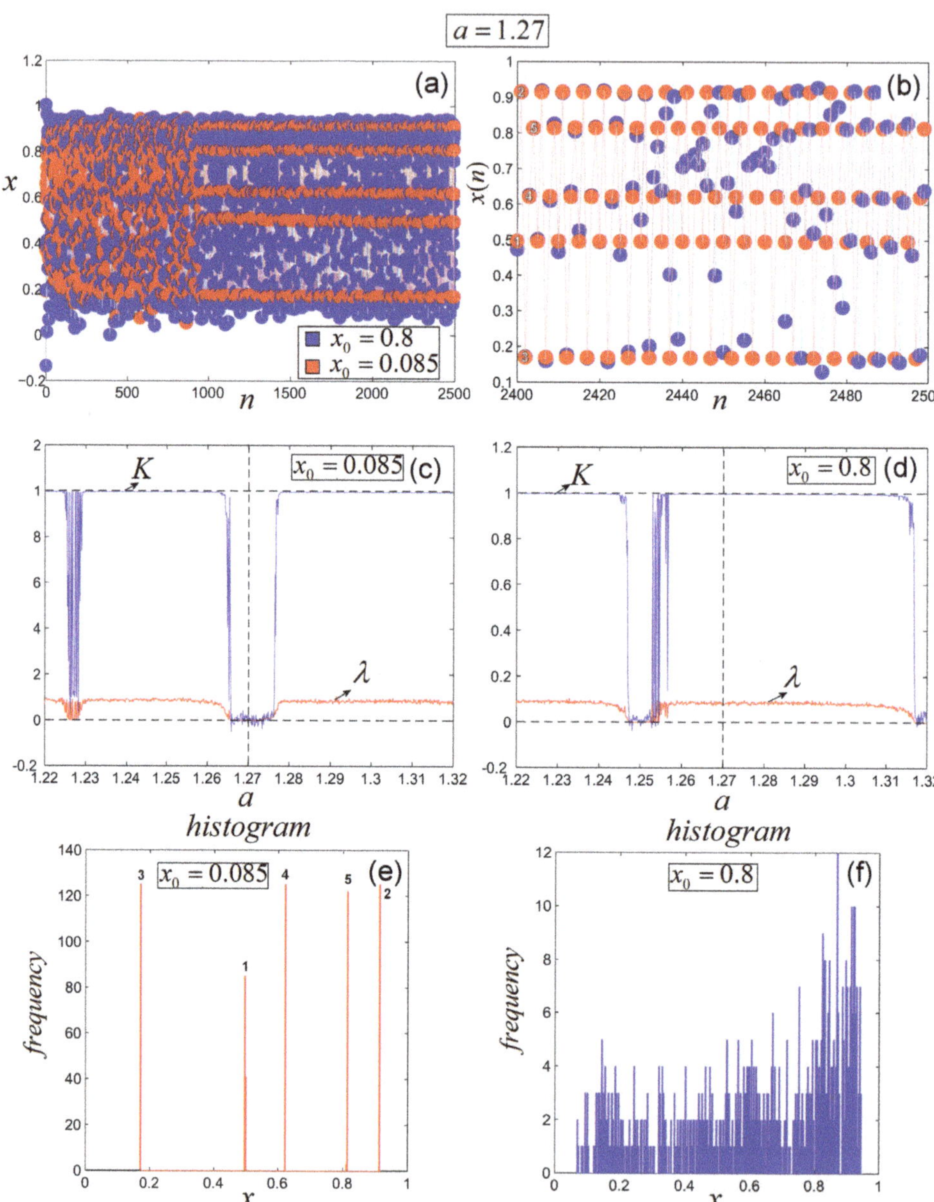

Figure 3. Attractors coexistence for $a = 1.27$ and $x_0 = 0.8$ and $x_0 = 0.085$. (**a**) superimposed time series, (**b**) Last 100 iterations of the time series; Numbered points indicate the CPO periodicity, (**c**) Asymptotic growth rate K and exponent λ for the attractor generated with $x_0 = 0.085$, (**d**) asymptotic growth rate K and exponent λ for the attractor generated with $x_0 = 0.8$, (**e**) histogram of the attractor generated with $x_0 = 0.085$; bars indicate the CPO elements revealed in (**b**), (**f**) histogram of the attractor generated with $x_0 = 0.085$.

3. Chaos Suppression in the FO Puu System

Being a model proposed in economy, where chaos is usually present, chaos suppression could represent an important task. A simple and efficient way used in this paper is to apply constant periodic impulses γ every Δ steps. The numerical form of the impulsive algorithm is

$$x(n+1) = \begin{cases} f(x(n)), & n \in \mathbb{N}, \\ (1+\gamma)x(n+1), & \text{if} \quad \bmod (n, \Delta) = 0, \end{cases} \tag{7}$$

where $f \in C(\mathbb{R}, \mathbb{R})$ (here the cubic map (1)), $\Delta \in \mathbb{N}^*$ and the impulse γ some relative small real number.

Numerically, the control algorithm (7) reads as follows: after the new value of x is calculated at the moment $n+1$, $x(n+1)$, if n is multiple of Δ steps, $x(n+1)$ is perturbed with $1+\gamma$.

The algorithm is suitable to systems where the state variable is accessible to perturbations and has been successfully utilized for discontinuous systems of FO [24], continuous fractional-order systems of IO [25], discrete systems of FO [26], and discrete systems of IO [27]. The stability of impulsive fractional difference equations is studied in [28]. Some analytical study, such as boundness and periodicity, that applied to a discrete economical supply and demand system can be found in [29].

Consider the case $q = 0.5$ and $a = 1.3$ when the both coexisting chaotic attractors of the Puu's system of FO are chaotic (see Figure 2b). This means that for whatever initial conditions x_0 the system evolves along one of the two chaotic attractors (red or blue).

To not modify the system structure, γ values are chosen to be relatively small, here of $1e-2$ order. The effectiveness of the algorithm (7) can be deduced form the bifurcation diagram of the controlled system vs. γ for $\gamma \in [-0.1, 0.1]$ and fixed Δ. The stable windows reveal the λ values generating stable CPOs. While the chaotic orbits are identified by the positiveness of the Lyapunov exponent, λ, the chaotic evolution of the underlying time series, dense set of bars in histograms, and $K \approx 1$, the COPs have negative (or zero) Lyapunov exponent, computationally periodic behavior in the time series, discrete bars in histograms and $K \approx 0$.

The case $\Delta = 1$, when the system is impulsed at every step, is presented in Figure 4a,d for x_0 chosen in the two mentioned attraction basins, $x_0 = 0.8$ and $x_0 = 0.085$, respectively. As can be seen in the bifurcation diagrams, the impulsive algorithm applied within the two attraction basins preserves the existence of the two previously existing chaotic attractors, but also generates stable CP windows in the γ. Obviously, it is desirable to find those γ values for which both coexisting chaotic attractors are stable. However, for $\Delta = 1$, the periodic windows cannot suppress simultaneously the chaos in both chaotic attractors (see dotted green and black lines). For example, for $\gamma = -0.0375$ (dotted black lines in Figure 4a,d), the chaotic attractor corresponding to $x_0 = 0.8$ of the impulsed system is stabilized and evolves along a stable CPO of period-5 (see the five intersections of the green dotted line in Figure 4a, the time series in Figure 4b and the history in Figure 4c), but the coexisting attractor corresponding to $x_0 = 0.085$ of the impulsed system is chaotic for the same value of γ (see Figure 4d–f). The stability within the periodic windows is revealed by the Lyapunov exponent (blue plot) and the asymptotic growth rate K (red plot) which are approximately 0.

Concluding, for $\Delta = 1$, it is possible that, depending on the initial conditions, the system evolves regularly but also chaotically. This impediment can be avoided if $\Delta = 2$, when the system is impulsed only every two steps. Now, both chaotic attractors can be controlled for a large interval of γ values (Figure 5). Thus, for $\gamma = -0.09$, the impulsed system evolves, for whatever initial condition, along a stable CPO: a stable CPO of period-8, for $x_0 = 0.8$ (Figure 5a–c), and also a stable CPO of period-4, for $x_0 = 0.085$ (Figure 5d–f).

Chaos suppression can also be realized with higher values of Δ. For example, for $\Delta = 3$ (Figure 6), if one sends an impulse to the system at every three steps, for $\gamma = -0.086$ (see dotted green line), and one obtains two stable CPOs of multiple periods.

Next, let the averaged energy of the impulsed system, $\bar{E}$, which can be determined with the following relation:

$$\bar{E} = \frac{\sum_{n=1}^{N} x^2(n)}{N},$$

for N sufficiently large (see e.g., [30] for the averaged energy applied to a generalized logistic map).

The variation of the averaged energy vs. γ, for the considered cases $\Delta = 1, 2, 3$, for $N = 2000$, is presented in Figure 7. Note that, if the system is impulsed rarer, i.e., $\Delta = 3$, the energy of the impulsed system (red) is smaller than that of the not impulsed system, $\bar{E}_0$ (horizontal dotted line) for most of γ values, while, if the system is impulsed more often, i.e., $\Delta = 1$ (black) or $\Delta = 2$ (blue), the necessary energy for the controlled system (vertical dotted lines) is bigger than $\bar{E}_0$. In addition, for positive values of γ (not considered in this work), $\bar{E} < \bar{E}_0$, for all considered values for Δ. As expected, when γ is negative, the system increases the energy.

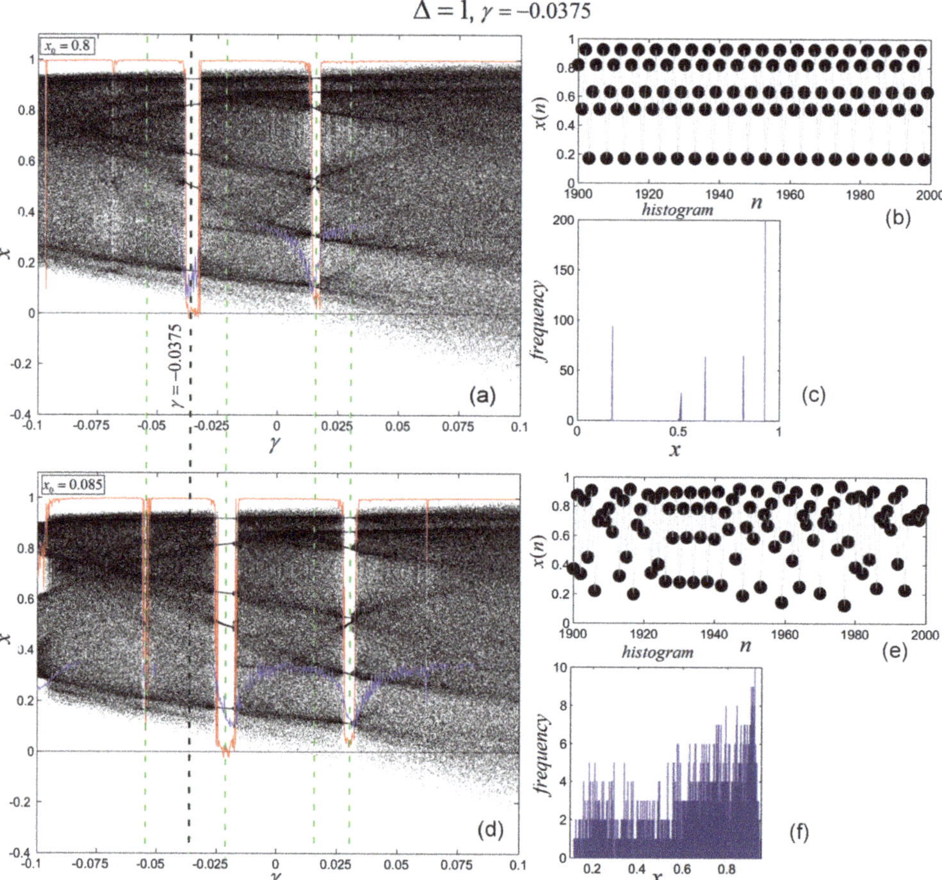

Figure 4. Impulsive algorithm applied for $\Delta = 1$ and $\gamma = -0.0375$ (dotted green line). (**a**,**d**) bifurcation diagrams of the impulsed system for $\gamma \in [-0.1, 0.1]$ and $x_0 = 0.8$ (top) and $x_0 = 0.085$ (bottom) respectively, (**b**,**e**) last 100 iterations of time series for $x_0 = 0.8$ and $x_0 = 0.085$, respectively, (**c**,**f**) histograms for for $x_0 = 0.8$ and $x_0 = 0.085$, respectively.

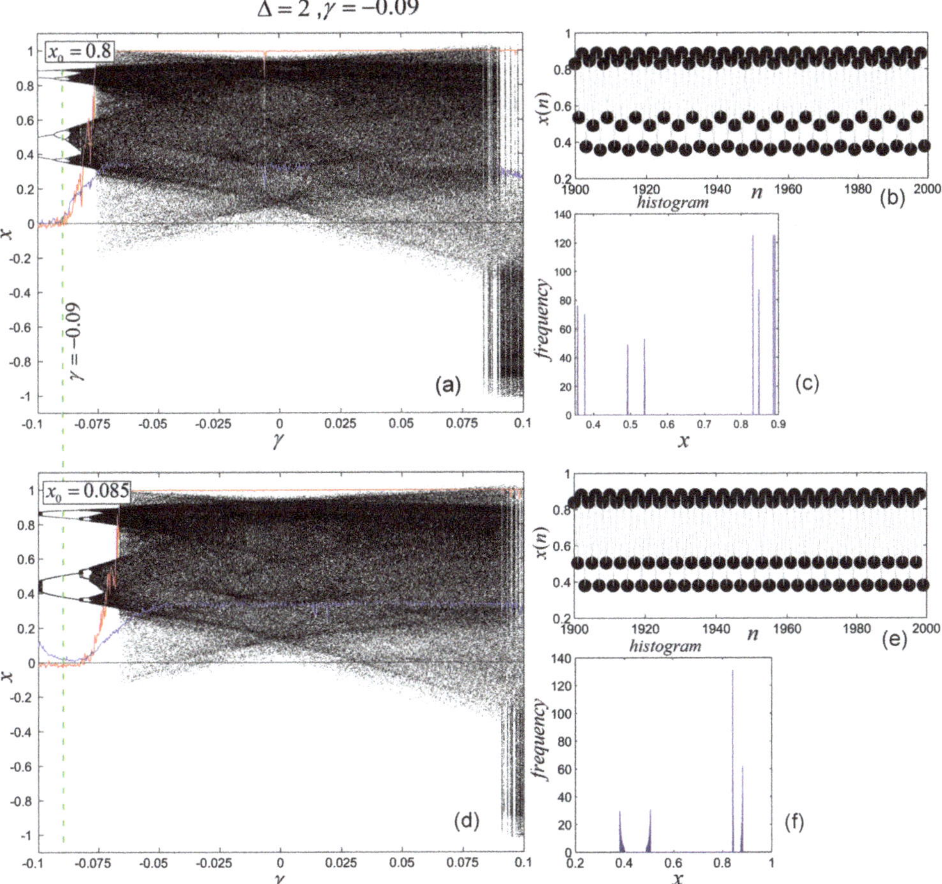

Figure 5. Chaos suppression with the impulsive algorithm for $\Delta = 2$ and $\gamma = -0.09$ (dotted green line). (**a**,**d**) bifurcation diagrams of the impulsed system for $\gamma \in [-0.1, 0.1]$ and $x_0 = 0.8$ (top) and $x_0 = 0.085$ (bottom), respectively, (**b**,**e**) last 100 iterations of time series for $x_0 = 0.8$ and $x_0 = 0.085$, respectively, (**c**,**f**) histograms for for $x_0 = 0.8$ and $x_0 = 0.085$, respectively.

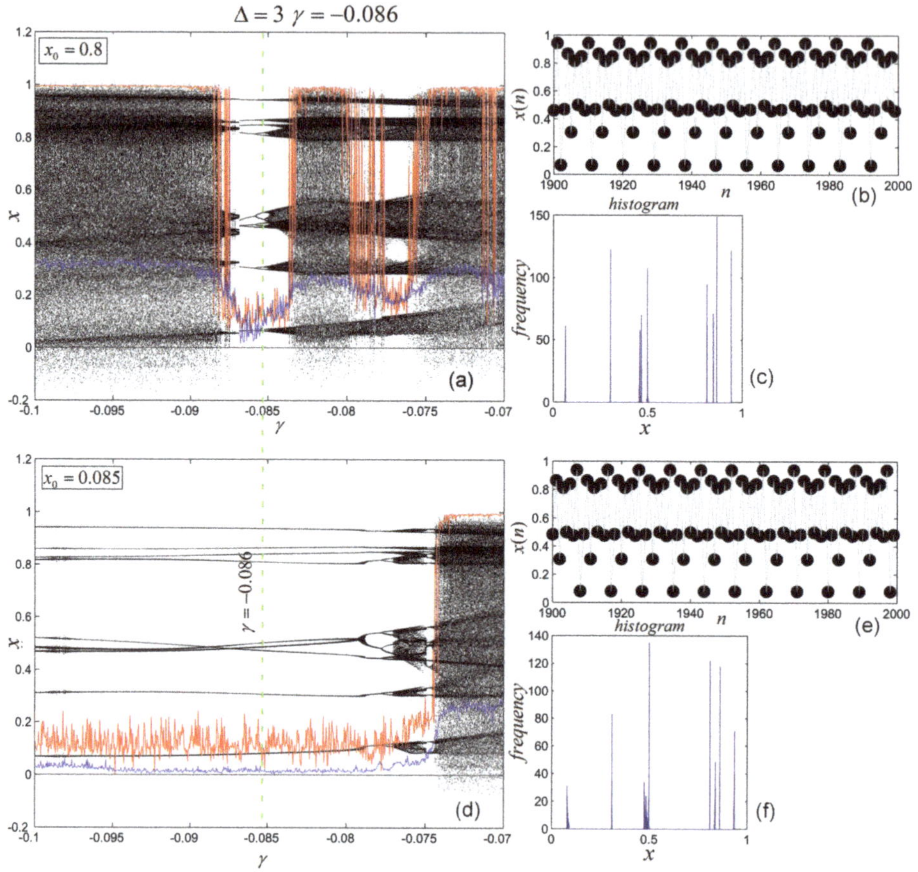

Figure 6. Chaos suppression for $\Delta = 3$ and $\gamma = -0.086$ (dotted green line). (**a**,**d**) bifurcation diagrams of the impulsed system for $\gamma \in [-0.1, 0.1]$ and $x_0 = 0.8$ (top) and $x_0 = 0.085$ (bottom), respectively, (**b**,**e**) last 100 iterations of time series for $x_0 = 0.8$ and $x_0 = 0.085$, respectively, (**c**,**f**) histograms for $x_0 = 0.8$ and $x_0 = 0.085$, respectively.

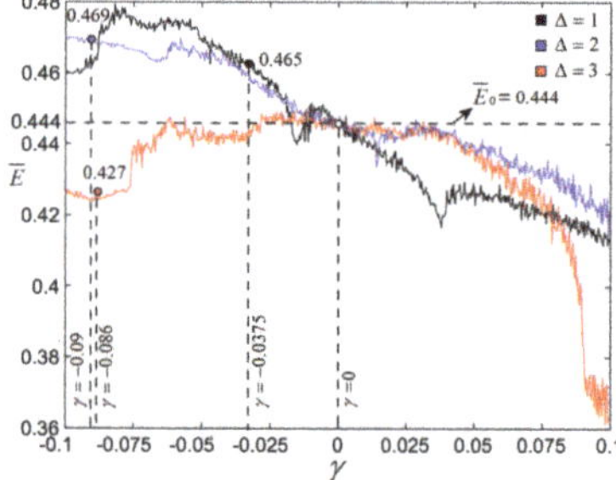

Figure 7. Superimposed averaged energy $\bar{E}$, for $\Delta = 1$ (black), $\Delta = 2$ (blue) and $\Delta = 3$ (red). The horizontal dotted line represents the average energy $\bar{E}_0$ of the not impulsed system.

4. Conclusions

In this paper, the FO variant of the Puu system has been considered and some of its properties are characterized comparatively with its IO counterpart. Contrary to the IO variant, the FO variant presents significant symmetries. In addition, contrary to the FO continuous-time systems, the chaos in the Puu system of FO reduces once the fractional order q increases.

Because, as usual for FO systems, the system does not admit periodic orbits, instead "periodic orbit", the notion of "computationally periodic orbit" has been introduced. In this way, the door of everything that involves "periodicity", such as stable/unstable cycles, chaos, chaos control, remains open for study. Note that another approach of this kind of orbits would be done via "almost periodicity" notion (see, e.g., [31]).

Using the impulsive algorithm (7), which perturbs periodically the state variable with the value $1+\gamma$, chaos can be suppressed. The values of γ can be determined from the bifurcation diagram of the impulsed system vs. γ. In this paper, the perturbations γ are negative, but, depending on the considered system, positive values can also be used (see the mentioned references). Note that, in nature, these kinds of perturbations can be shocks, and can appear in natural disasters, ecology, or can be used in ecosystems management, harvesting, etc.

As for other examples of nonlinear discrete FO systems modeled by Caputo's derivative, chaos in Puu's system of FO vanishes when the bifurcation parameter approaches 1. Therefore, the problem of chaos suppression makes sense only for relatively smaller values of the fractional order q, where the considered system behaves chaotically.

Funding: This research received no external funding.

Conflicts of Interest: The author declares no conflict of interest.

References

1. Lakshmikantham, V.; Bainov, D.; Simeonov, P. *Theory of Impulsive Differential Equations*; World Scientific Publishing Co. Pte. Ltd.: Singapore, 1989.
2. Yang, T. *Impulsive Control Theory*; Springer: Berlin/Heidelberg, Germany, 2001.
3. Bonotto, E.M.; Bortolan, M.C.; Caraballo, T.; Collegari, R. A survey on impulsivedynamical systems. *Electron. J. Qual. Theory* **2016**, *7*, 1–27.
4. Rozhko, V.F. A certain class of almost periodic motions in systems with pulses. *Differ. Uravn.* **1972**, *8*, 2012–2022. (In Russian)
5. Kaul, S.K. On impulsive semidynamical systems. *J. Math. Anal. Appl.* **1990**, *150*, 120–128. [CrossRef]
6. Oldham, K.B.; Spanier, J. The Fractional Calculus: Theory and Applications of Differentiation and Integration to Arbitrary Order. In *Mathematics in Science and Engineering*; Academic Press: New York, NY, USA, 1974; Volume 11.
7. Podlubny, I. Fractional Differential Equations. In *Mathematics in Science and Engineering*; Academic Press: San Diego, CA, USA, 1999; Volume 198.
8. Diaz, J.B.; Olser, T.J. Differences of fractional order. *Math. Comput.* **1974**, *28*, 185–202. [CrossRef]
9. Tenreiro Machado, J.A.; Silva, M.F.; Barbosa, R.S.; Jesus, I.S.; Reis, C.M.; Marcos, M.G.; Galhano, A.F. Some applications of fractional calculus in engineering. *Math. Probl. Eng.* **2010**, *2010*, 639801. [CrossRef]
10. Samuelson, P.A. Interactions between the multiplier analysisand the principle of acceleration. *Rev. Econ. Stat.* **1939**, *21*, 75–78. [CrossRef]
11. Hicks, J.R. *A Contribution to the Theory of the Trade Cycle*; Oxford University Press: Oxford, UK, 1950.
12. Puu, T.; Sushko, I. A business cycle model with cubic nonlinearity. *Chaos Soliton. Fract.* **2004**, *19*, 597–612. [CrossRef]
13. Rogers, T.; Whitley, D.C. Chaos in the cubic mapping. *Math. Model.* **1983**, *4*, 9–25. [CrossRef]
14. Chossat, P.; Golubitsky, M. Iterates of maps with symmetry. *SIAM J. Math. Anal.* **1988**, *19*, 1259–1270. [CrossRef]
15. King, G.; Stewart, I. Symmetric Chaos. *Math. Sci. Eng.* **1992**, *185*, 257–315. [CrossRef]

16. Wu, G.-C.; Baleanu, D. Discrete fractional logistic map and its chaos. *Nonlinear Dyn.* **2014**, *75*, 283–287. [CrossRef]
17. Xin, B.; Liu, L.; Hou, G.; Ma, Y. Chaos Synchronization of Nonlinear Fractional Discrete Dynamical Systems via Linear Control. *Entropy* **2017**, *19*, 351. [CrossRef]
18. Goodrich, C.; Peterson, A.C. *Discrete Fractional Calculus*; Springer: Berlin/Heidelberg, Germany, 2015.
19. Wu, G.-C.; Baleanu, D. Jacobian matrix algorithm for Lyapunov exponents of the discrete fractional maps. *Commun. Nonlinear Sci. Numer. Simulat.* **2015**, *22*, 95–100. [CrossRef]
20. Tavazoei, M.S.; Haeri, M. A proof for non existence of periodic solutions in time invariant fractional order systems. *Automatica* **2009**, *45*, 1886–1890. [CrossRef]
21. Diblik, J.; Feckan, M.; Pospisil, M. Nonexistence of periodic solutions and S-asymptotically periodic solutions in fractional difference equations. *Appl. Math. Comput.* **2015**, *257*, 230–240. [CrossRef]
22. Cuevas, C.; Souza, J. S-asymptotically ω-periodic solutions of semilinear fractional integro-differential equations. *Appl. Math. Lett.* **2009**, *22*, 865–870. [CrossRef]
23. Gottwald, G.; Melbourne, I. A new test for chaos in deterministic systems. *Proc. R. Soc. A* **2004**, *460*, 603–611. [CrossRef]
24. Danca, M.-F.; Garrappa, R. Suppressing chaos in discontinuous systems of fractional order by active control. *Appl. Math. Comput.* **2015**, *257*, 89–102. [CrossRef]
25. Danca, M.-F.; Tang, W.K.S.; Wang, Q.; Chen, G. Suppressing chaos in fractional-order systems by periodic perturbations on system variables. *Eur. Phys. J. B* **2013**, *86*, 79. [CrossRef]
26. Danca, M.-F.; Feckan, M.; Kuznetsov, N. Chaos control in the fractional order logistic map via impulses. *Nonlinear Dyn.* **2019**, *98*, 1219–1230. [CrossRef]
27. Danca, M.-F.; Codreanu, S.; Bako, B. Detailed analysis of a nonlinear prey-predator model. *J. Biol. Phys.* **1997**, *23*, 11–20. [CrossRef] [PubMed]
28. Wu, G.-C.; Baleanu, D. Stability analysis of impulsive fractional difference equations. *Fract. Calc. Appl. Anal.* **2018**, *21*, 354–375. [CrossRef]
29. Danca, M.-F.; Feckan, M. Hidden chaotic attractors and chaos suppression in an impulsivediscrete economical supply and demand dynamical system. *Commun. Nonlinear Sci.* **2019**, *74*, 1–13. [CrossRef]
30. Stavroulaki, M.; Sotiropoulos, D. The Energy of Generalized Logistic Maps at Full Chaos. *Chaotic Model. Simul.* **2012**, *3*, 543–550.
31. Akhmet, M. *Almost Periodicity in Chaos*. In *Almost Periodicity, Chaos, and Asymptotic Equivalence*; Springer: Cham, Switzerland, 2020; Volume 27.

Article

Dynamical Properties of Fractional-Order Memristor

Shao Fu Wang [1,2] and Aiqin Ye [1,*]

1 College of Electrical and Electronic Engineering, Anhui Science and Technology University, Bengbu 233100, China; ncdx11@126.com

2 College of Electronic and information Engineering, NanJing University of Aeronautics and Astronautics, NanJing 211106, China

* Correspondence: yeaq@ahstu.edu.cn

Received: 12 January 2020; Accepted: 27 February 2020; Published: 9 March 2020

Abstract: The properties of a fractional-order memristor is studied, and the influences of parameters are analyzed and compared. The results reflect that the resistance value of a fractional-order memristor can be affected by fraction-order, frequency, the switch resistor ratio, average mobility and so on. In addition, the circuit of a fractional-order memristor that is serially connected and connected in parallel with inductance and capacitance are studied. Then, the current–voltage characteristics of a simple series one-port circuits that are composed of a fractional-order memristor and a capacitor, or composed of a fractional-order memristor and a inductor are studied separately. The results demonstrate that at the periodic excitation, the memristor in the series circuits will have capacitive properties or inductive properties as the fractional order changes, the dynamical properties can be used in a memristive circuit.

Keywords: fraction-order memristor; device property; serial and parallel; Dynamical analysis

---

1. Introduction

Fractional calculus, an important branch of mathematics, was born in 1695 and appeared almost simultaneously to classic calculus. Fractional calculus, in a narrow sense, mainly includes fractional differentials and fractional integrals, and it broadly includes fractional differences and fractional sum quotients. Since the theory of fractional calculus has been successfully applied to various fields in recent years, people have gradually discovered that fractional calculus can describe some non-classical phenomena in the fields of natural science and engineering applications. The current popular areas of fractional calculus include fractional numerical algorithms and fractional synchronization.

The successful development of a memristor has provided a new avenue for electronic technology and information technology, and it is expected to realize new functions. Their non-volatility makes memristors play a key role in memory, neural networks, and pattern recognition. Chua firstly defined a memristor (MR) in 1971 [1], and HP Labs reported the successful fabrication of nanoscale memristive devices [2]. A memristor is the fourth two-terminal fundamental circuit element with information storage ability and, as such, has attracted immense worldwide interest from both industry and academia [3]. Applications of MRs are used in many fields such as filter design, programmable logic, biological systems, and neural systems such as neural synaptic weighting with a pulse-based memristor circuit [4], Boolean logic operations and computing circuits based on memristors [5], and the voltage–current relationship of active memristors and frequency [6]. A generalized boundary condition memristor model was proposed in [7]. Research on a coupling behavior-based series-parallel flux-controlled memristor has been also conducted [8–10]. Two types of nanoscale nonlinear memristor models and their series-parallel circuit have been investigated [11], as have the characteristics of a memristor and its application in the circuit design [12–17]. First order mem-circuits have been studied [18]. Research on the equivalent analysis circuit of a memristors network [19], memristor-based adaptive coupling for consensus and synchronization [20–22], and coupling as a third relation in memristive systems have been proposed [23–30]. However, there is less research on fractional-order memristor, and there has been no literature physical background. In this paper, the analytical solution of parameter expression is derived by using a fractional-order memristor model in Section 2. The properties of fractional-order memristor is studied in Section 3, The results reflect that the resistance value of a fractional-order memristor can be affected by fraction-order, frequency, the switch resistor ratio, average mobility, and so on; additionally, the circuit of M^{α} serially connected and connected in parallel with M^{α}, L and C are also studied separately. Finally, the conclusion is presented in Section 4.

2. The Fractional Derivative

The α order Caputo derivative is defined as:

$$D_t^{\alpha} x(t) = \frac{1}{\Gamma(m-a)} \int_0^t \frac{x^{(m)}(\tau)}{(t-\tau)^{1+\alpha-m}} d\tau. \tag{1}$$

where $m = [\alpha] + 1$ and $\Gamma(m)$ is the Euler's gamma function; when $\alpha \in (0,1)$:

$$D_t^{\alpha} e^{\lambda t} = \lambda^{\alpha} e^{\lambda t}. \tag{2}$$

Then, we obtain:

$$D_t^{\alpha} e^{i\omega t} = (i\omega)^{\alpha} e^{i\omega t}. \tag{3}$$

The real part and imaginary part of the sine and cosine functions can be obtained by separating Equation (3).

Some basic properties of Caputo fractional calculus are as follows.

(1) ${}_aD_t^{\alpha}[uf(t) + vg(t)] = u_a D_t^{\alpha} f(t) + v_a D_t^{\alpha} g(t)$
(2) ${}_aD_t^{\alpha}{}_aD_t^{\beta} f(t) = {}_aD_t^{\beta}{}_aD_t^{\alpha} f(t) = {}_aD_t^{\alpha+\beta} f(t)$,
(3) ${}_a^LD_t^{\alpha} K = \frac{Kt^{-\alpha}}{\Gamma(1-\alpha)}$, $\alpha > 0$

In which the constant $K \neq 0$, and the Caputo FD is ${}_a^CD_t^{\alpha} K = 0$, $\alpha > 0$.

3. The Model of Fractional-Order Memristor

The structure of an MR and its symbol are shown in Figure 1.

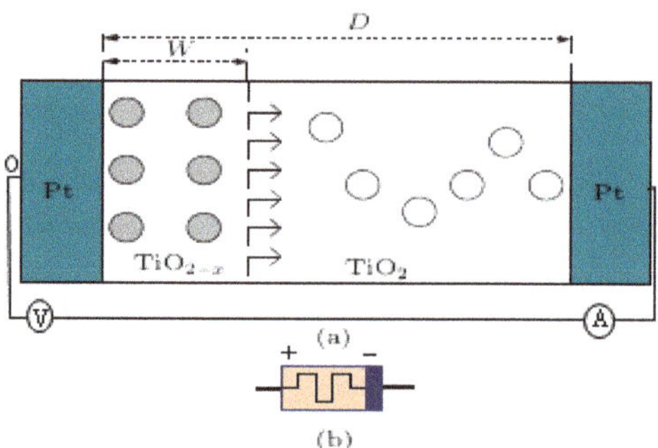

Figure 1. The structure and symbol of a memristor (MR). (**a**) The structure and (**b**) the symbol.

The mathematical model are defined as follows

$$v(t) = M^{\alpha}(x) i(t). \tag{4a}$$

$$M^{\alpha}(x) = R_{ON}x + (1-x)R_{OFF}. \tag{4b}$$

$$x(t) = k_0 D_t^{-\alpha} i(t). \tag{4c}$$

In which $D_t^{-\alpha}$ is the α integral of $x(t)$, $M^{\alpha}(x)$ is a fractional-order memristor when supposing the input current $i(t) = I_m \sin(\omega t)$, switch resistor $R_{ON} = 100\ \Omega$ and $R_{OFF} = 10\ \text{k}\Omega$, the current $I_m = 0.2$ mA $D = 10$ nm, the average mobility $u_v = 10^{-14}$ m.s^{-1}V^{-1}, the length $x_0 = 0.01$, and the frequency $\omega = 5$ rad/s. The transient current curve and the voltage curve of the memristor are shown in Figure 2a,b respectively.

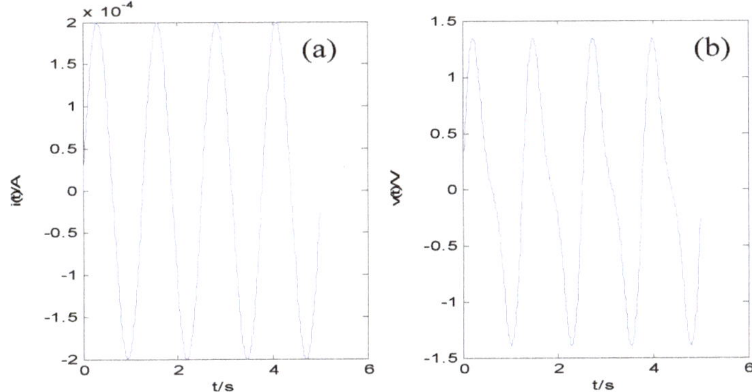

Figure 2. (**a**) The transient current curve and (**b**) the transient voltage curve.

The simulation result for when α has different values is shown in Figure 3a. It was found the an MR possesses memristive properties, with pinched hysteresis loops forming inclined "8", and this property can be used to realize signal storage or computing. From Figure 3b, it can be seen that the smaller the value of fractional order, the greater the dynamic range amplitude of the resistance was.

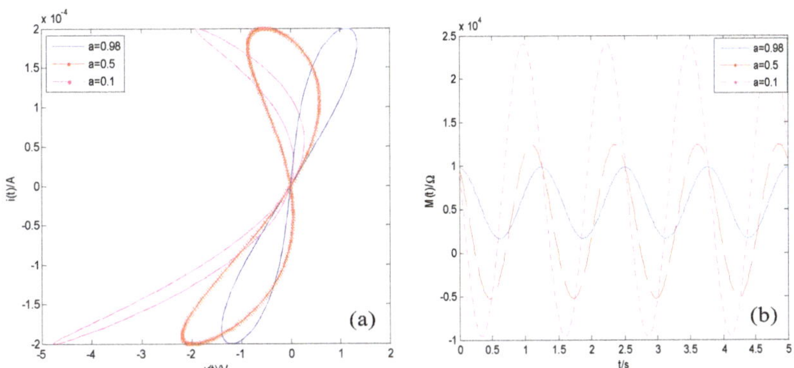

Figure 3. (**a**) The curves of $v-i$ and (**b**) the curves of $M(t)$.

The characteristic curves of a fractional-order memristor with different parameters are shown in Figure 4. From Figure 4a, it can be seen that the difference of resistance decreased as the frequency ω increased. From Figure 4b, it can be seen that the switch resistor increased and the curves of $v-i$ were inclined to right. From Figure 4c, it can be seen that the difference of resistance increased as the value of μ_v increased.

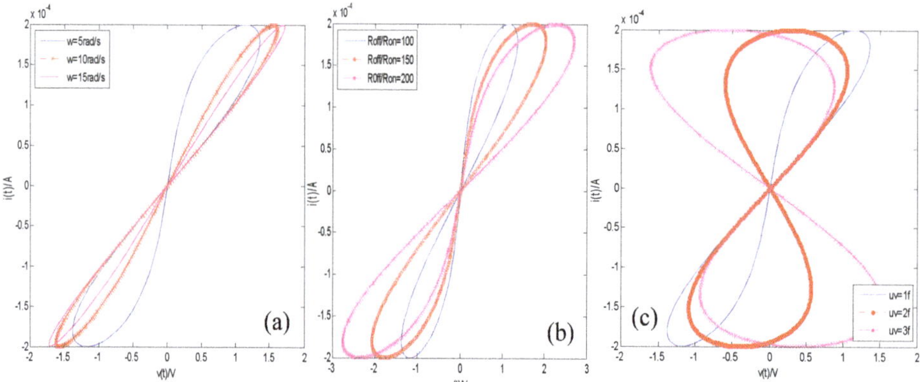

Figure 4. The $v - i$ curves of a fractional-order MR ($\alpha = 0.98$). (**a**) Varied with frequency ω; (**b**) varied with different switch resistance (R_{OFF}/R_{ON}); and (**c**) varied with different average mobility values μ_v.

4. The Properties of Fractional-Order Memristor

The fractional-order memristor in six cases of connection, serially and in parallel, are discussed in this section.

4.1. The Two Fractional-Order Memristors in Serial

The serial circuit of two fractional-order memristors is shown in Figure 5.

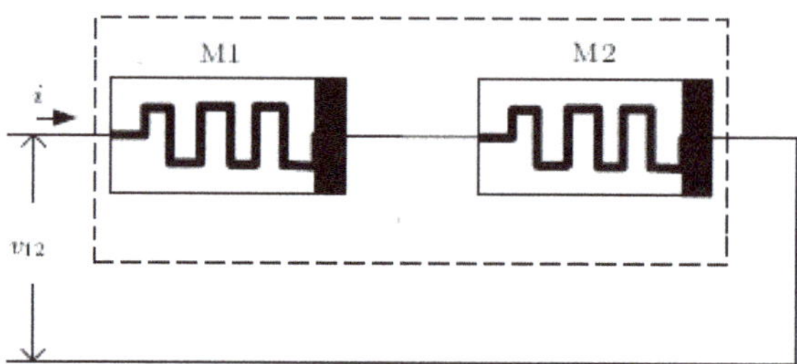

Figure 5. Serial circuit of two fractional-order memristors.

By choosing the current $i(t) = I_m \sin(\omega t)$, based on the Caputo differential, Euler's formulas, and the separating variables method, we obtain:

$$\dot{x}(t) = kD_t^{1-\alpha} i(t) = kI[D_t^{1-\alpha} \sin \omega t]. \tag{5}$$

When $t >> 1$, Equation (11) can be simplified as:

$$\dot{x}(t) \approx kI\omega^{1-\alpha} \sin(\omega t + \frac{1-\alpha}{2}\pi). \tag{6}$$

Then, the two sides of Equation (12) can be integrated to get:

$$x(t) \approx x(0) + \frac{kI}{\omega^{\alpha}}[\cos(\frac{1-\alpha}{2}\pi) - \cos(\omega t + \frac{1-\alpha}{2}\pi)]. \tag{7}$$

By choosing the parameters M1($u_v = 10^{-14}$ m.s^{-1}V^{-1}), M2($u_v = 2 \times 10^{-14}$ m.s^{-1}V^{-1}), and $\alpha = 0.98$, one can obtain the curves of $v - i$ that are are shown in Figure 6.

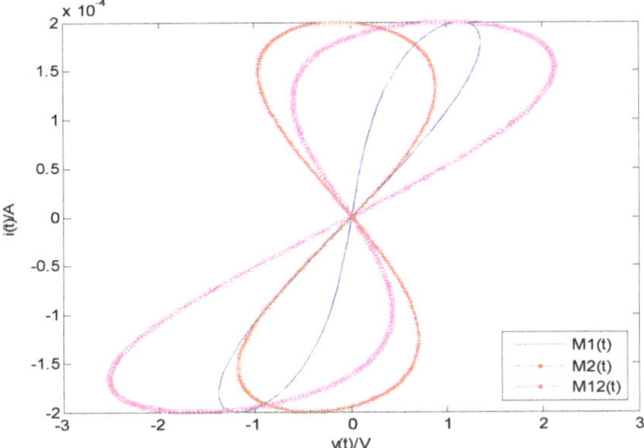

Figure 6. Simulation results of two fractional-order, serially connected memristors $v(t) - i(t)$ curves.

It was found the two serial MRs also possessed the memristive properties with pinched hysteresis loops behaving as inclined "8", as shown in Figure 6. It can be seen that the values of M1, M2 and M12 are increased with the value of v(t), and the value of M12 is bigger than those of M1 and M2.

4.2. The Circuit of Fractional-Order MR in Parallel

The circuit of two fractional-order MRs in parallel is shown in Figure 7. According to Equations (4b) and (4c), it be written as:

$$M_\alpha(x)\dot{x}(t) = k_0 D_t^{-\alpha} M_\alpha(x) i(t). \tag{8a}$$

$$M_\alpha(x)\dot{x}(t) = k_0 D_t^{-\alpha} v(t). \tag{8b}$$

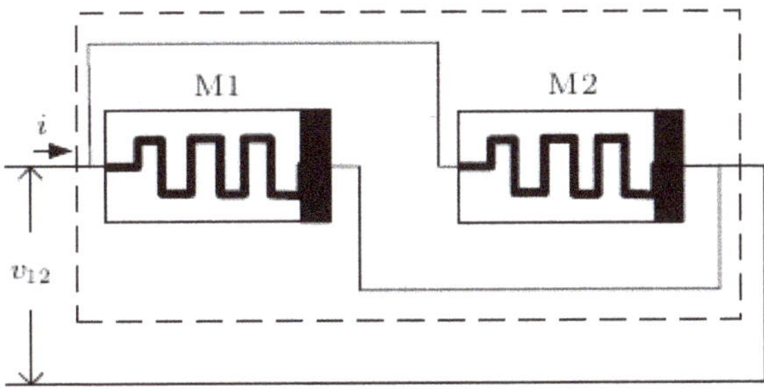

Figure 7. Circuit of two fractional-order memristors n parallel.

By choosing the voltage $v(t) = V_m \sin(\omega t)$, one can obtain:

$$M_\alpha(x)\dot{x}(t) = k_0 V_m D_t^{1-\alpha} \sin(\omega t)\dot{x}(t) = k D_t^{1-\alpha} i(t) = kI[D_t^{1-\alpha} \sin \omega t]. \tag{9}$$

When $t >> 1$,Equation (11) can be simplified as:

$$M_\alpha(x)\dot{x}(t) \approx k_0 V_m \omega^{1-\alpha} \sin(\omega t + \frac{1-\alpha}{2}\pi). \tag{10}$$

and

$$[R_{ON}x + R_{OFF}(1-x)] \cdot (x - x_0) \approx \frac{kV_m}{\omega^\alpha}[\cos(\frac{1-\alpha}{2}\pi) - \cos(\omega t + \frac{1-\alpha}{2}\pi)]. \tag{11}$$

$$(R_{ON} - R_{OFF})x^2 + [(1 + x_0)R_{OFF} - R_{ON}x_0]x = \frac{kV_m}{\omega^\alpha}[\cos(\frac{1-\alpha}{2}\pi) - \cos(\omega t + \frac{1-\alpha}{2}\pi)] + R_{OFF}x_0. \quad (12)$$

Then, one can set $A = R_{ON} - R_{OFF}$; $B = (1 + x_0)R_{OFF} - R_{ON}x_0$; $H = -\frac{kV_m}{\omega^\alpha}[\cos(\frac{1-\alpha}{2}\pi) - \cos(\omega t + \frac{1-\alpha}{2}\pi)] + R_{OFF}x_0$
and we can obtain:

$$Ax^2 + Bx + H = 0. \quad (13)$$

$$x_1 = \frac{-B + \sqrt{B^2 - 4AH}}{2A}; x_2 = \frac{-B - \sqrt{B^2 - 4AH}}{2A}. \quad (14)$$

The value of $M_\alpha(x)$ can be calculated.

To further study the dynamic behaviors of this MR circuit in parallel, the parameters were configured as the following: the input voltage $v(t) = V_m \sin \omega t$, and the parameters $R_{ON1} = 100\ \Omega$, $R_{OFF1} = 10\ \text{k}\Omega$, $V_m = 3$ V, $D = 10$ nm, $u_v = 10^{-14}$ m.s^{-1}V^{-1}, $R_{ON2} = 120\ \Omega$, $R_{OFF2} = 18\ \text{k}\Omega$, and ω = 5. The simulation result when using these parameters is shown in Figure 8.

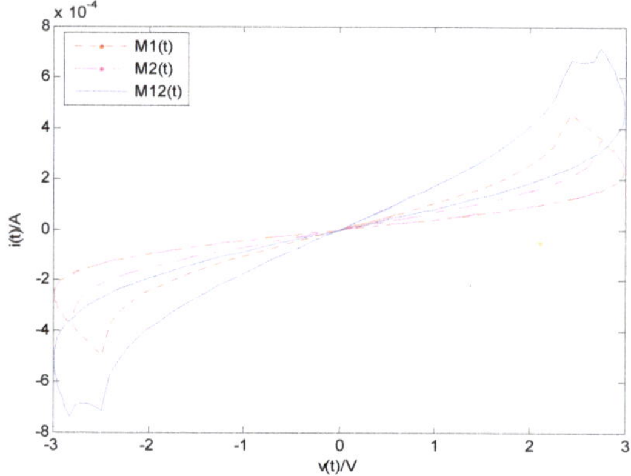

Figure 8. Simulation results of two memristors connected in parallel with $v(t) - i(t)$ curves.

It was found the two MRs that were connected in parallel also possessed the memristive properties with pinched hysteresis loops which are hown in Figure 8.

4.3. The Circuit of Fractal-Order Memristor and Capacitor That Are Serially Connected

The fractional-order memristor $M^\alpha(t)$ and serially connected capacitor C are shown in Figure 9. By assume the current $i(t) = I_m \sin \omega t$, M expresses the memristor, and C is the capacitor.

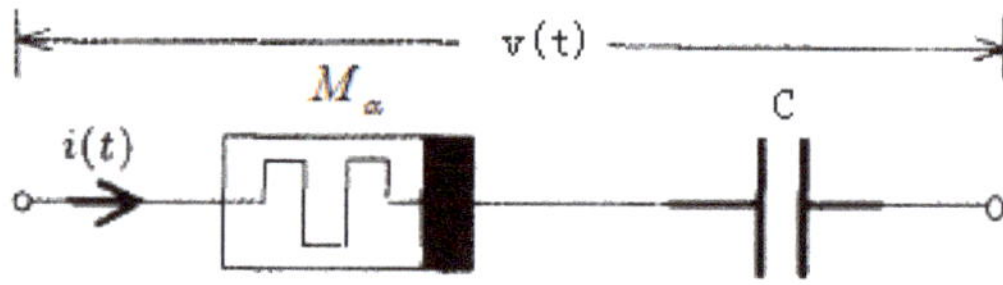

Figure 9. The circuit of the fractional-order memristor and connected serially capacitor.

We can obtain:

$$\begin{aligned} v(t) &= u_C(t) + u_{M_\alpha}(t) = u_C(t_0) + \tfrac{1}{C}\int_{t_0}^{t} i_C(t)dt + i_{M_\alpha}(t)M_\alpha(t) \\ &= u_C(t_0) + \tfrac{I\cos(\omega t_0)}{C\omega} - \tfrac{I\cos(\omega t)}{C\omega} - B\sin(\tfrac{1-\alpha}{2}\pi) + A\sin(\omega t) + B\sin(2\omega t + \tfrac{1-\alpha}{2}\pi). \end{aligned} \quad (15)$$

where

$$A = (R_{ON} - R_{OFF})[x(0) + \frac{Ik}{\omega^{\alpha}}(\cos\frac{1-\alpha}{2}\pi)]I + R_{OFF}I. \tag{16a}$$

$$B = \frac{(R_{OFF} - R_{ON})kI^2}{2\omega^{\alpha}}. \tag{16b}$$

The simulation result is shown in Figure 10. From Figure 10a, it can be seen as the parameter α decreased, the area of hysteresis loops increased and the difference of resistance increased with same current. From Figure 10b, it can be seen that if the parameters α and C were not varied, as ω increased, the area of hysteresis loops decreased. From Figure 10c, it can be seen that if the parameters α and ω were not varied, the area of hysteresis loops was not almost changed and the capacitor has little effect on the circuit.

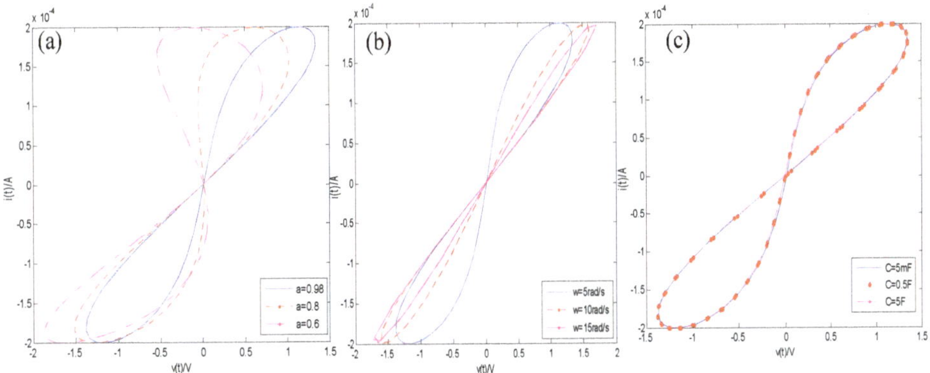

Figure 10. $M^{\alpha}C$ series circuit and its features. (**a**) Effect of the order α; (**b**) effect of the exciting frequency ω; and (**c**) effect of the capacitance C.

4.4. The Circuit of Fractal-Order Memristor and Capacitor That Were Connected in Parallel

A memristor and a capacitor which were connected in parallel are shown in Figure 11, we can assume that $v(t) = V_m \sin(\omega t)$, $\omega(0) = 0$,V_m is the voltage magnitude.

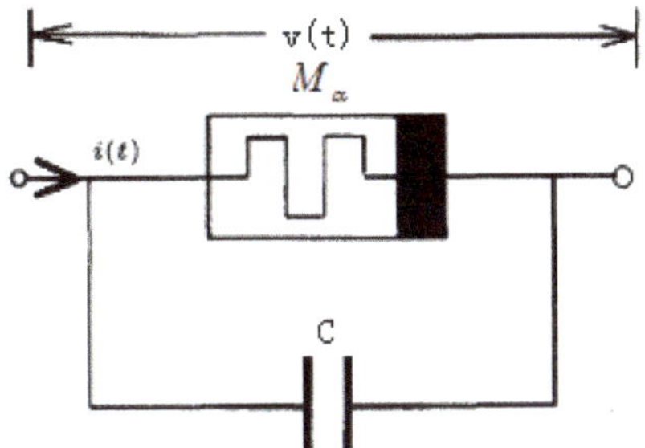

Figure 11. The circuit of a memristor and a capacitor that were connected in parallel.

By applying Kirchhoff's current law (KCL), the current of circuit can be written as:

$$i(t) = \frac{v(t)}{M^{\alpha}(t)} + C\frac{dv(t)}{dt} = \frac{v(t)}{M^{\alpha}(t)} + \omega C V_m \cos(\omega t). \tag{17}$$

Additionally, when the parameters α and ω were chosen as different values, it can be seen the circuit of a memristor and a capacitor that were connected in parallel also possessed the memristive properties with pinched hysteresis loops which are shown in simulations in Figure 12a,b.

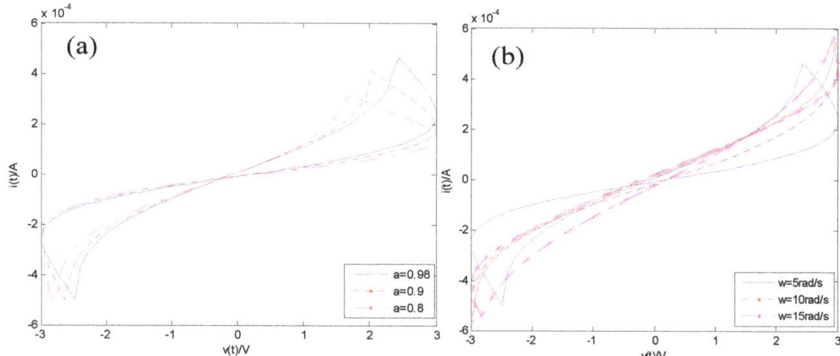

Figure 12. The response of the circuit of a memristor and a capacitor that were connected in parallel. (**a**) $\alpha = 0.98, 0.9$, and 0.8; and (**b**) $\omega = 5, 10, 15$ rad/s.

4.5. The Circuit of Fractal-Order Memristor and Inductor That Are Serially Connected

The circuit of the memristor and serially connected inductor are shown in Figure 13, where we assumed $i(t) = I_m \sin(\omega t)$, $\omega(0) = 0$.

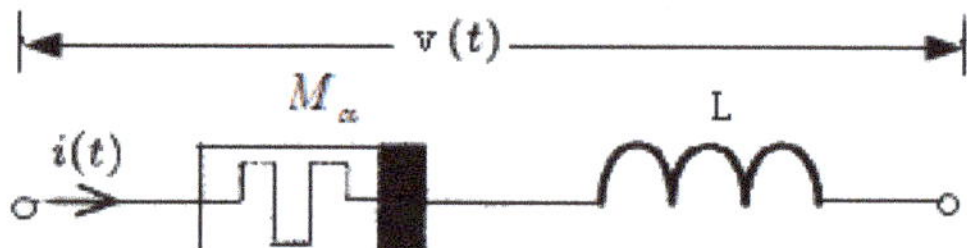

Figure 13. The circuit of the memristor and serially connected inductor.

The voltage was:

$$\begin{aligned} v(t) = u_L(t) + u_{M_\alpha}(t) = L\tfrac{di_L(t)}{dt} + i_{M_\alpha}(t)M_\alpha(t) \\ = IL\omega\cos(\omega t) + A\sin(\omega t) + B[\sin(2\omega t + \tfrac{1-\alpha}{2}\pi) - \sin(\tfrac{1-\alpha}{2}\pi)] \\ = u_C(t_0) + \tfrac{I\cos(\omega t_0)}{C\omega} - \tfrac{I\cos(\omega t)}{C\omega} - B\sin(\tfrac{1-\alpha}{2}\pi) + A\sin(\omega t) + B\sin(2\omega t + \tfrac{1-\alpha}{2}\pi). \end{aligned} \quad (18)$$

and

$$A = (R_{ON} - R_{OFF})[x(0) + \frac{Ik}{\omega^\alpha}(\cos\frac{1-\alpha}{2}\pi)]I + R_{OFF}I. \quad (19a)$$

$$B = \frac{(R_{OFF} - R_{ON})kI^2}{2\omega^\alpha}. \quad (19b)$$

The simulation result is shown in Figure 14. From Figure 14a, it can be seen that as the parameters ω and L were not varied, because as α decreased, the area of hysteresis loops increased and the difference of resistance increased with the same current. From Figure 14b, it can be seen that if the parameters α and L were not varied, as ω increased, the area of the hysteresis loops decreased. From Figure 14c, it can be seen that if the parameters α and ω were not varied, the area of the hysteresis loops was not almost changed, and it is shown that the inductor barely affected the circuit.

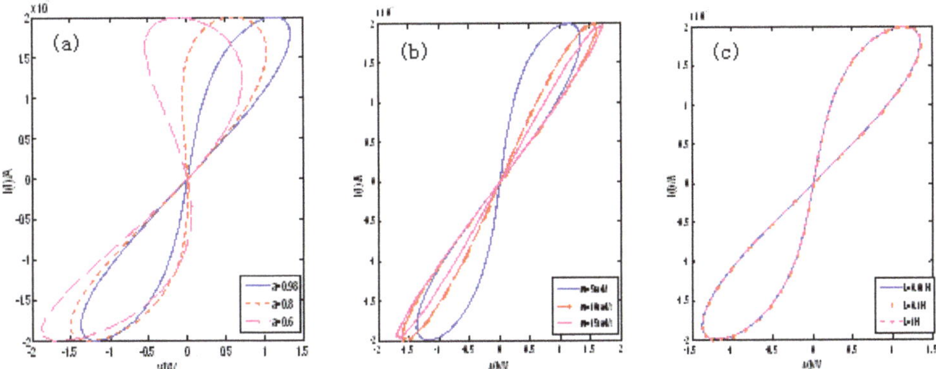

Figure 14. $M^{\alpha}L$ series circuit and its features: (**a**) Effect of α on the hysteresis loop; (**b**) effect of ω on the hysteresis loop; and (**c**) effect of L on the hysteresis loop.

4.6. The Circuit of Fractal-Order Memristor and Inductor Connected in Parallel

A memristor and a capacitor that were connected in parallel are shown in Figure 15. By assuming the voltage $v(t) = V_m \sin(\omega t)$, $\omega(0) = 0$, V_m can be found as the voltage amplitude. By applying Kirchhoff's current law (KCL), the current of circuit can be written as:

$$L\frac{di_L(t)}{dt} = v(t) = V_m \sin(\omega t). \tag{20a}$$

$$i_L(t) = \frac{V_m}{\omega L}[1 - \cos(\omega t)]. \tag{20b}$$

$$i(t) = i_{M_\alpha}(t) + i_L(t) = \frac{v(t)}{M_\alpha(t)} + \frac{V_m}{\omega L}[1 - \cos(\omega t)]. \tag{20c}$$

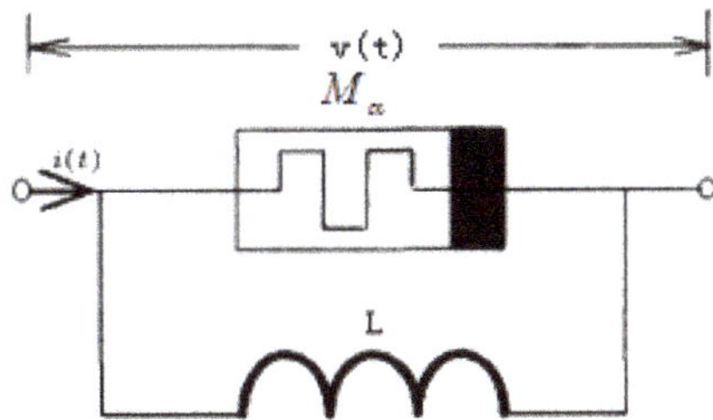

Figure 15. The circuit of a memristor and a inductor that were connected in parallel.

The simulation is shown in Figure 16a,b. It can be seen the circuit of the memristor and inductor that were connected in parallel also possessed memristive properties with pinched hysteresis loops. From Figure 16a, it can be seen the area of pinched hysteresis loops increased with α increased. From Figure 16b, it can be seen the area of the pinched hysteresis loops decreased with the ω increased.

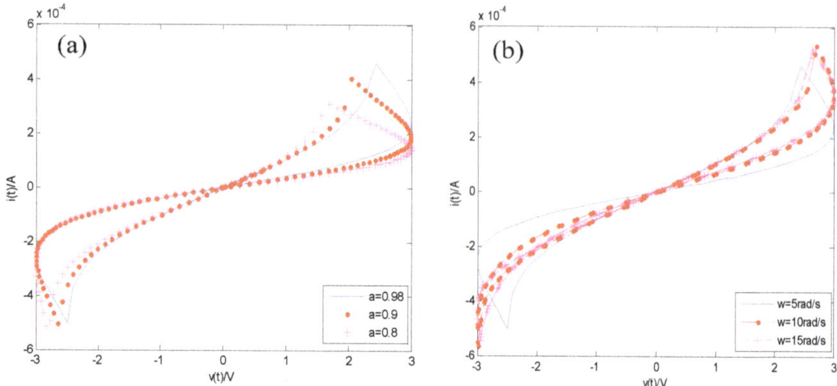

Figure 16. The response of the circuit of the memristor and inductor connected in parallel. (**a**) $\alpha = 0.98$, 0.9, and 0.8; and (**b**) $\omega = 5, 10, 15$ rad/s.

5. Conclusions

In summary, this paper has presented a fractional-order memristor model and verifies the three essential characteristics of a fractional-order memristor. In addition, the properties of fractional-order memristor have been described. In a simple memristive series circuit, with a change of the fractional derivative order, the series circuit of a fractional-order memristor and a capacitor or inductor shows a conversion from a pure capacitor circuit to a memristive circuit. The series circuit shows conversions of purely inductive and memristive circuits. Here, analytical solutions were derived by a fractional-order memristor, the properties of fractional-order memristor model parameters were obtained, and simulation results were given. The results showed that material properties determine the order of the fractional derivative, so the best memory capacity of a physical memristor can be achieved by finding materials that are compatible with the excitation frequency.

Author Contributions: Conceptualization, S.F.W. and A.Y.; methodology, S.F.W; software, S.F.W.; validation, S.F.W. and A.Y.; formal analysis, S.F.W.; investigation, S.F.W.; resources, S.F.W.; data curation, S.F.W.; writing—original draft preparation, S.F.W.; writing—review and editing, S.F.W.; visualization, S.F.W.; supervision, S.F.W.; project administration, S.F.W.; funding acquisition, A.Y. All authors have read and agreed to the published version of the manuscript.

Funding: This work was supported by the Foundation of The University Synergy Innovation Program of Anhui Province under Grant GXXT-2019-019 and Key R&D Projects of Anhui Science and Technology Department under Grand 201904f06020022.

Conflicts of Interest: The authors declare that there is no conflict of interest regarding the publication of this paper.

References

1. Chua, L.O. Memristor—The missing circuit element. *IEEE Trans. Circuit Theory* **1971**, *CT-18*, 507–519. [CrossRef]
2. Strukov, D.B.; Snider, G.S.; Stewart, D.R.; Williams, R.S. The missing memristor found. *Nature* **2008**, *453*, 80–83. [CrossRef] [PubMed]
3. Yu, D.; Iu HH, C.; Liang, Y.; Fernando, T.; Chua, L.O. Dynamic Behavior of Coupled Memristor Circuits. *IEEE Trans. Circuits Syst. I Regul. Pap.* **2015**, *62*, 1607–1615. [CrossRef]
4. Kim, H.; Sah, M.P.; Yang, C.; Roska, T.; Chua, L.O. Neural synaptic weighting with a pulse-based memristor circuit. *IEEE Trans. Circuit Syst. I Reg. Pap.* **2012**, *59*, 148–158. [CrossRef]
5. Papandroulidakis, G.; Vourkas, I.; Vasileiadis, N.; Sirakoulis, G.C. Boolean logic operations and computing circuits based on memristors. *IEEE Trans. Circuits Syst. II Exp. Briefs.* **2014**, *61*, 972–976. [CrossRef]

6. Bao, B.-C.; Zou, X. Voltage-Current Relationship of Active Memristor and Frequency Characteristic of Active Memristive Circuit. *Acta Electron. Sin.* **2013**, *41*, 593–596.
7. Ascoli, A.; Corinto, F.; Tetzlaff, R. Generalized boundary condition memristor model. *Int. J. Circuit Theor. Appl.* **2015**, *44*, 60–84. [CrossRef]
8. Wang, Y.; Yang, J. Research of coupling behavior based series-parallel flux-controlled memristor. *Acta Phys. Sin.* **2015**, *64*, 237303.
9. Sozen, H. First-order memristor–capacitor filter circuits employing hp memristor. *J. Circuits Syst. Comput.* **2014**, *23*, 1450116. [CrossRef]
10. Song, D.-H.; Lü, M.-F.; Ren, X.; Li, M.-M.; Zu, Y.-X. Basic properties and applications of the memristor circuit. *Acta Phys. Sin.* **2012**, *61*, 118101.
11. Dong, Z.-K.; Duan, S.-K. Two types of nanoscale nonlinear memristor models and their series-parallel circuit. *Acta Phys. Sin.* **2014**, *63*, 128502.
12. Wang, Q.-H.; Song, W.-P. Characteristics of memristor and it's application in the circuit design. *Electron. Compon. Mater.* **2014**, *33*, 5–7.
13. Frasca, M.; Gambuzza, L.V. Implementation of adaptive coupling through memristor. *Phys. Status Solidi* **2015**, *12*, 206–210. [CrossRef]
14. Gambuzza, L.-V.; Buscarino, A. Memristor-Based Adaptive Coupling for Consensus and Synchronization. *IEEE Trans. Circuits Syst. I Regul. Pap.* **2015**, *62*, 1175–1184. [CrossRef]
15. Wang, T.-S.; Zhang, R.-D. Properties of memristor in RLC circuit and diode circuit. *Acta Phys. Sin.* **2014**, *63*, 178101.
16. Duan, S.-K.; Hu, X.-F. Memristor-based RRAM with applications. *Sci. China* **2012**, *42*, 754–768. [CrossRef]
17. Yu, Y.J.; Wang, Z.H. A fractional-order memristor model and the fingerprint of the simple series circuits including a fractional-order memristor. *Acta Phys. Sin.* **2015**, *64*, 238401.
18. Riaza, R. First Order Mem-Circuits: Modeling, Nonlinear Oscillations and Bifurcations. *IEEE Trans. Circuits Syst. I Regul. Pap.* **2013**, *60*, 1570–1582. [CrossRef]
19. He, B.-X.; Bao, B.-C. Research on the Equivalent Analysis Circuit of Memristors Network and Its Characteristics. *J. Electron. Inf. Technol.* **2012**, *34*, 1252–1256.
20. Yuan, F.; Wang, G.-Y. Study on dynamical characteristics of a meminductor model and its meminductor-based oscillator. *Acta Phys. Sin.* **2015**, *64*, 210504.
21. Xu, H.; Tian, X.B. Influence of temperature change on conductive characteristics of titanium oxide memristor. *Acta Phys. Sin.* **2014**, *63*, 098402.
22. Papoutsidkis, M.; Kalovrektis, K.; Drosos, C. Design of an Autonomous Robotic vehicle for area mapping and remote monitroing. *Int. J. Comput. Appl.* **2017**, *12*, 36–41.
23. Cai, W.; Tetzlaff, R. Beyond series and parallel: Coupling as a third relation in memristive systems. In Proceedings of the 2014 IEEE International Symposium on Circuits and Systems (ISCAS), Melbourne, VIC, Australia, 1–5 June 2014; pp. 1259–1262.
24. Long, Z.X.; Zhang, Y. Noether's theorem for fractional variational problem from El-Nabulsi extended exponentially fractional integral in phase space. *Acta Mech.* **2014**, *225*, 77–90. [CrossRef]
25. El-Nabulsi, A.R. Fractional derivatives generalization of Einstein's field equations. *Indian J. Phys.* **2013**, *87*, 195–200. [CrossRef]
26. Fellah, Z.; Depollier, C.; Fellah, M. Application of fractional calculus to the sound waves propagation in rigid porus materials:validation via ultrasonic measurements. *Acta Acust. United Acust.* **2002**, *88*, 34–39.
27. Magin, R.; Ortigueira, M.D.; Podlubny, I.; Trujillo, J. On the fractional signals and systems. *Signal. Process.* **2011**, *91*, 350–371. [CrossRef]
28. Ostalczyk, P. *Discrete Fractional Calculus: Applications in Control and Image Processing*; World Scientific: Singapore, 2015.
29. Das, S. *Functional Fractional Calculus*; Springer Science & Business Media: Berlin, Germany, 2011.
30. Kilbas, A.A.; Marichev, O.I.; Samko, S.G. *Fractional Integrals and Derivatives, Theory and Applications*; Gordon and Breach: Yverdon, Switzerland, 1993.

Article

Fractional Levy Stable and Maximum Lyapunov Exponent for Wind Speed Prediction

Shouwu Duan [1], Wanqing Song [1], Carlo Cattani [2,3,*], Yakufu Yasen [4] and He Liu [1]

1 School of Electronic & Electrical Engineering, Shanghai University of Science Engineering, Shanghai 201620, China; 15070238383@163.com (S.D.); swqls@126.com (W.S.); nh324310@163.com (H.L.)
2 Engineering School, DEIM, University of Tuscia, 01100 Viterbo, Italy
3 Department of Mathematics and Informatics, Azerbaijan University, 1007 Baku, Azerbaijan
4 State Grid Kashi Electric Power Supply Company, 156 Renmin West Road, Kashi City 844099, China; 18299639290@163.com
* Correspondence: cattani@unitus.it

Received: 11 March 2020; Accepted: 2 April 2020; Published: 11 April 2020

Abstract: In this paper, a wind speed prediction method was proposed based on the maximum Lyapunov exponent (Le) and the fractional Levy stable motion (fLsm) iterative prediction model. First, the calculation of the maximum prediction steps was introduced based on the maximum Le. The maximum prediction steps could provide the prediction steps for subsequent prediction models. Secondly, the fLsm iterative prediction model was established by stochastic differential. Meanwhile, the parameters of the fLsm iterative prediction model were obtained by rescaled range analysis and novel characteristic function methods, thereby obtaining a wind speed prediction model. Finally, in order to reduce the error in the parameter estimation of the prediction model, we adopted the method of weighted wind speed data. The wind speed prediction model in this paper was compared with GA-BP neural network and the results of wind speed prediction proved the effectiveness of the method that is proposed in this paper. In particular, fLsm has long-range dependence (LRD) characteristics and identified LRD by estimating self-similarity index H and characteristic index α. Compared with fractional Brownian motion, fLsm can describe the LRD process more flexibly. However, the two parameters are not independent because the LRD condition relates them by $\alpha H > 1$.

Keywords: wind speed forecasting; fractional Levy stable motion; long-range dependence; Lyapunov exponent

1. Introduction

When the penetration of wind power exceeds a certain value, it seriously affects power quality. At present, the error rate of wind speed forecasting of wind farms is about 25%–40%, and the research on wind speed forecasting of wind farms has not reached a satisfactory level [1]. If wind speed and wind power could be accurately predicted, it would be beneficial for the power system dispatching department to adjust the scheduling plan in time, which could effectively reduce the impact of wind power on the power grid [2]. At the same time, the improvement of prediction accuracy could also reduce the operating cost and rotation reserve of power systems [3], increase the limit of wind power penetration, and lay the foundation for wind farms to participate in bidding for power generation [4]. Many researchers have developed several different wind speed prediction methods. The simplest prediction method is the continuous method, which uses the closer wind speed or power observation value as the prediction value for the next point [5]. Other prediction methods include Kalman filters [6], ARMA [7], artificial neural network (ANN) [8], fuzzy logic, and so on. These methods only need the wind speed or power time series of the wind farm to build a model and make predictions. The spatial correlation method needs to consider the wind farm and the wind speed time series of several places close to it, using several locations. Then, the spatial correlation between wind speeds is used to predict the wind speed of a wind farm and to predict wind power.

In recent years, the prediction of stochastic sequences with long-range dependence (LRD) characteristics has become a hot topic and can be applied to the prediction of non-stationary stochastic processes. The LRD model [9,10] can give better forecasting of the stochastic sequence by comprehensively considering the influence of both the past state and the current state on the future state. Fractional Brownian motion models with LRD characteristics have been widely applied in this field [11–13]. The LRD of fractional Brownian motion is described by the only parameter H (self-similarity index). Compared with fractional Brownian motion the LRD of the fractional Levy stable motion (fLsm) is determined instead by two parameters α and H, which can separately characterize the local irregularity and global persistence [14] so that fLsm can describe the long correlation process more flexibly. Therefore, in the following, we used a prediction model of stochastic sequences based on fLsm with LRD to predict wind speed.

The prediction method used in this paper involves the maximum Lyapunov exponent and fLsm iterative prediction model [15]. The Lyapunov exponent can help us distinguish between noise and signals that obey a certain law. In this paper, we mainly used the reciprocal of the maximum Lyapunov exponent to represent the maximum predictive steps. The methods for calculating the Lyapunov exponent are the definition method, small-data method, wolf method, Jacobian method, etc. This paper used the small-data method [16], which makes full use of all available data and therefore has relatively high accuracy. The small-data method is fast in operation and easy to implement, and it shows strong robustness to the embedding dimension and time delay, as well as the size of the data amount. However, the choice of the embedding dimension is subjective, and the time delay is not necessarily accurate. Therefore, we needed to use the c-c method [17,18] to avoid this problem.

The fLsm iterative prediction model was established by fLsm-driven Langevin-type stochastic differential equation (SDE) [19]. First, the fractional Black-Scholes model [20,21] was extended and the parameterized SDE was obtained. Then, the fLsm was discretized by Taylor series expansion of fractional order [22], and the mathematical relationship between the increment of flsm and Levy's stable white noise was obtained and substituted into discrete Langevin-type SDE. Finally, using the discrete Langevin-type SDE and the difference equation, the expression of the proposed fLsm finite difference iterative prediction model was obtained.

Wind speed is mainly affected by weather and terrain shape, and it also changes with altitude, so randomness is the basic property of wind speed, at least on a small scale. In this paper, we used Langevin-type SDE [19] driven by fLsm to describe the randomness of wind speed. However, the wind speed data in most regions do not have heavy tail characteristics, which will lead to a larger error when using wind speed data to estimate the parameters of the fLsm iterative prediction model. From the characteristics of the data: weighting can increase the variance of the data so that the data can show heavy-tailed features.

This paper is organized as follows. The small-data method is introduced in Section 2, The fLsm is introduced in Section 3, where we also analyze the model and LRD characteristics. The fLsm finite difference iterative forecasting model is proposed in Section 4, which establishes the finite-difference iterative forecasting model by making Langevin-type SDE [19] driven by fractional Levy stable motion, and the Langevin-type SDE [19] parameters estimated by the novel Characteristic Function (CF) method [23–25]. The wind speed forecasting results show the superiority of the method used in this paper (Section 5). The mathematical relationship between wind speed and wind power is introduced in Section 6. Concluding remarks are given in Section 7.

2. Maximum Prediction Steps Based on Lyapunov Exponent

The small-data method [16] is defined as follows.

Let $\{x_1, x_2 \cdots x_N\}$, be a given chaotic time series, then the reconstructed phase space is defined as:

$$Y_i = \left(x_i, x_{i+\tau,} \cdots, x_{i+(m-1)\tau}\right) \in R^m, (i = 1, 2, \cdots, M), \tag{1}$$

where $N = M + (m-1)\tau$. The embedding dimension m and the time delay τ can be chosen according to the C-C method [17,18].

After the reconstruction of phase space, find the nearest adjacent point of each point on the given orbit, i.e.,

$$d_j(0) = \min_{x_{\hat{j}}} \|Y_j - Y_{\hat{j}}\|, \tag{2}$$

$$\left|j - \hat{j}\right| > p, \tag{3}$$

where p is the average period of the time series, which can be estimated by the inverse of the average frequency of the power spectrum, and the maximum Le can be estimated by the average divergence rate of each point on the basic orbit. For each reference point, calculate the distance to the nearest discrete point after the first discrete time step by

$$d_j(i) = \min_{x_{\hat{j}}} \|Y_{j+i} - Y_{\hat{j}+i}\|,\ i = 1, 2, \cdots, \min\left(M - j, M - \hat{j}\right), \tag{4}$$

The average divergence rate obeys the exponential divergence, i.e.,:

$$d_j(i) = C_j e^{\lambda_1 (i \cdot \Delta t)},\ C_j = d_j(0), \tag{5}$$

Take the logarithm on both sides to get:

$$ln\ d_j(i) = lnC_j + \lambda_1 (i \cdot \Delta t), \tag{6}$$

Obviously, the maximum Le is roughly equivalent to the slope on this set of straight lines. It can be obtained by approximating this set of lines by the method of least squares.

$$\lambda_1 = \frac{lnd_j(i) - lnC_j}{i \cdot \Delta t}, \tag{7}$$

The reciprocal of the maximum Lyapunov exponent is the maximum prediction steps ε when $\lambda_1 > 0$.

$$\varepsilon = \frac{1}{\lambda_1}, \tag{8}$$

3. Fractional Levy Stable Motion

3.1. Parameter Meaning of Levy Stable Motion

Levy stable motion represents a non-Gaussian random process with LRD and high variability, we denote by $X \sim S_\alpha(\beta, \delta, \mu)$ the stable distribution with parameters α, β, δ and μ. Its characteristic function form is as follows [26]:

$$\varphi(\theta : \alpha, \beta, \delta, \mu) = E\left[e^{j\theta x}\right] = \begin{cases} \exp\left\{j\mu\theta - \delta|\theta|^\alpha\left[1 - j\beta\frac{\theta}{|\theta|}\tan\left(\frac{\pi\alpha}{2}\right)\right]\right\}, \alpha \neq 1 \\ \exp\left\{j\mu\theta - \delta|\theta|^\alpha\left[1 + j\beta\frac{\theta}{|\theta|}\frac{2}{\pi}\ln|\theta|\right]\right\}, \alpha = 1 \end{cases}, \tag{9}$$

where $\delta > 0$, $\beta \in [-1, 1]$, $\mu \in R$. The parameter β is called the skewness parameter, while δ is called scale parameter, and μ the location parameter. In this article, we studied symmetric stable distribution, so we make $\beta = 0$. The location parameter μ indicates the mean, and the scale parameter δ represents the discrete nature of the distribution.

Where $\alpha \in (0,2]$. The parameter α is the tail parameter and the distribution is Gaussian when $\alpha = 2$, whereas the tail is exponential. In what follows, we typically supposed $0 < \alpha < 2$. When $x \to \infty$, the probability tails of X satisfy [27]:

$$P\{|X| > x\} \sim C_a \delta^{\alpha} x^{-\alpha}, \tag{10}$$

where C_a is a constant. The tail of the distribution with $0 < \alpha < 2$ obeys a power law and decreases to zero so slowly that the variance is infinite; the smaller the value of α, the slower the decrease. From the perspective of probability distribution, as the value of α decreases, its tail becomes thicker (Figure 1).

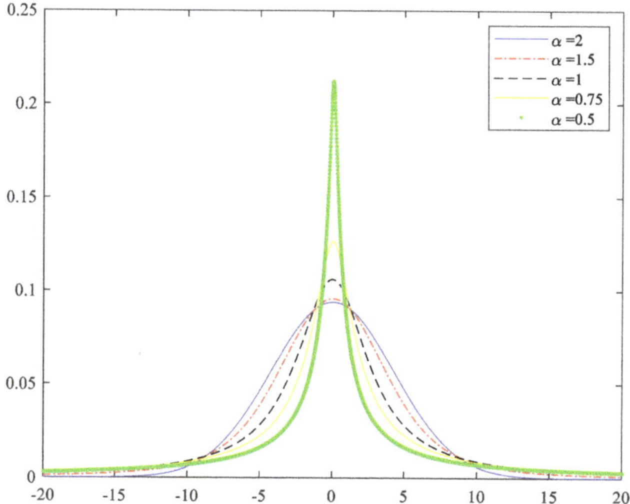

Figure 1. Influence of different characteristic index values on the probability distribution function.

3.2. Long-Range Dependence and Self-Similarity Fractional Levy Stable Motion

The model of fLsm [14] is given by the following stochastic integral:

$$L_{H,\alpha}(t) = \int_{-\infty}^{\infty} \left\{ a\left[(t-s)_{+}^{H-\frac{1}{\alpha}} - (-s)_{+}^{H-\frac{1}{\alpha}} \right] + b\left[(t-s)_{-}^{H-\frac{1}{\alpha}} - (-s)_{-}^{H-\frac{1}{\alpha}} \right] \right\} Mds, \tag{11}$$

where a and b are the arbitrary constants, $x_{+}^{H-1/\alpha} = 0$ for $x \le 0$ and $x_{+}^{H-\frac{1}{\alpha}} = x^{H-\frac{1}{\alpha}}$ for $x > 0$, $M \in R$ is the symmetric Levy stable random measure, and H is the self-similarity parameter. The incremental process of fLsm [28] is as follows:

$$\begin{aligned} X_{H,\alpha}(t) &= L_{H,\alpha}(t+1) - L_{H,\alpha}(t) \\ &= \int_{-\infty}^{\infty} \{ a\left[(t+1-s)_{+}^{H-1/\alpha} - (-s)_{+}^{H-1/\alpha} \right] \\ &+ b\left[(t+1-s)_{-}^{H-1/\alpha} - (-s)_{-}^{H-1/\alpha} \right] \} \omega_{\alpha}(s), \end{aligned} \tag{12}$$

where $\omega_{\alpha}(s)$ is the Levy stable white noise.

Symmetric Levy stable motion is $1/\alpha$ self-similar, namely, $L_{\alpha}(t) \triangleq a^{-1/\alpha} L_{\alpha}(at)$ for all $a > 0$. Laskin et al. [29] have shown that the fLsm is a self-similar process with self-similar parameter $H - 1/2 + 1/\alpha$. The incremental process $\left\{ L_{H,\alpha}(t_2) - L_{H,\alpha}(t_1) \right\}$ is also self-similar with $H - 1/2 + 1/\alpha$.

The key parameters α, H of the fLsm model are not independent in some cases, i.e., the fLsm has LRD characteristics for $\alpha H > 1$ [30]. It is worth noting that the fLsm model has no long memory when $0 < \alpha < 1$, therefore, the range of α is limited to $(1, 2)$ to ensure that the fLsm model has the LRD characteristic. At the same time, $0.5 < H < 1$ is also required.

4. Iterative Forecasting Model Based on Fractional Levy Stable Motion

4.1. Iterative Forecasting Model

Let us consider the following Langevin-type stochastic differential equation driven by Levy stable motion [19]:

$$dX(t) = b(t, X(t))dt + \delta(t, X(t))dL_{\alpha}(t),\ X(0) = X_0, \tag{13}$$

where $dL_{\alpha}(t)$ stands for the increments of Levy α-stable motion $L_{\alpha}(t)$. By replacing $L_{H,\alpha}(t)$ to $L_{\alpha}(t)$, we obtain the Langevin-type stochastic differential equation driven by fractional Levy stable motion:

$$dX_{H,\alpha}(t) = b(t, X_{H,\alpha}(t))dt + \delta(t, X_{H,\alpha}(t)), dL_{H,\alpha}(t)X_{H,\alpha}(0) = X_0, \tag{14}$$

where $b(t, X(t))$ and $\delta(t, X(t))$ represent the drift and diffusion functions, respectively.

The fractional Black-Scholes model [20,21], which was developed by W. DAI et al. [31,32] has expression in the form:

$$dS_t = \mu S_t dt + \delta S_t dB_H(t), \tag{15}$$

where μ indicates the expected return rate and δ is the volatility rate. The Levy stable distribution is the Gaussian distribution when $\alpha = 2$ so that when $\alpha = 2$ the fLsm becomes the fractional Brownian motion, μ represents the mean, and δ represents the diffusion coefficient. The parameters b and δ in the Levy stable distribution represent the mean and diffusion coefficient, respectively, in $1 < \alpha \leq 2$. Consequently, Equation (14) can be rewritten as follows:

$$dX_{H,\alpha}(t) = \mu X_{H,\alpha}(t)dt + \delta X_{H,\alpha}(t)dL_{H,\alpha}(t), \tag{16}$$

where μ and δ are constants. They are derived from the novel CF method in the Appendix.

By using the Maruyama symbol [22], $dB_t = w(t)(dt)^{1/2}$, the following equations can be obtained:

$$\int_0^t f(\tau)(d\tau)^a = \rho \int_0^\tau (t-\tau)^{a-1} f(\tau)d\tau, \tag{17}$$

$$dx = f(t)(dt)^a, \tag{18}$$

where $0 < a < 1$, and a represents the self-similar parameter of x. The incremental expression of fLsm can be obtained by replacing $f(t)$ with $w_{\alpha}(t)$:

$$dL_{H,\alpha} = w_{\alpha}(t)(dt)^{H-\frac{1}{2}+\frac{1}{\alpha}}, \tag{19}$$

Equation (16) can be written the discrete form, which reads as follows:

$$\Delta X_{H,\alpha}(t) = \mu X_{H,\alpha}(t)\Delta t + \delta X_{H,\alpha}(t)w_{\alpha}(t)(\Delta t)^{H-\frac{1}{2}-\frac{1}{\alpha}}, \tag{20}$$

The iterative predictive model was obtained from the identity $\Delta X(t) = X(t+1) - X(t)$:

$$L_{H,\alpha}(t+1) = L_{H,\alpha}(t) + \mu L_{H,\alpha}(t)\Delta t + \delta L_{H,\alpha}(t)w_{\alpha}(t)(\Delta t)^{H-\frac{1}{2}+\frac{1}{\alpha}}, \tag{21}$$

4.2. Parameter Estimation with the Characteristic Function

In the essay of Wang et al. [23–25], some methods were introduced and the validity of these methods was compared, including the quantiles method, empirical characteristic function method, logarithmic moment method, Monte Carlo method, etc. It was concluded that the CF accuracy method was better. The parameter estimation methodology can be subdivided into the following steps:

Step 1: Let $x_i|_{i=1...N}$ be the sampling data for the fLsm,

Step 2: δ estimation:

$$\left|\varphi(\theta;\alpha,\beta,\mu,\delta)\right| = \left|E\left\{e^{j\theta x}\right\}\right| = e^{-\gamma|\theta|^{\alpha}}, \tag{22}$$

$$\ln\left|\varphi(\theta;\alpha,\beta,\mu,\delta)\right| = -\gamma|\theta|^{\alpha}, \tag{23}$$

$$\delta = -\ln\left|\varphi(1;\alpha,\beta,\mu,\delta)\right| = -\ln\left|E\left\{e^{jx}\right\}\right|, \tag{24}$$

The estimated δ has the form:

$$\hat{\delta} = -\ln\left|\hat{\varphi}(1;\alpha,\beta,\mu,\delta)\right| = -\ln\frac{1}{N}\left|\sum_{i=1}^{N} e^{jx_i}\right|, \tag{25}$$

Step 3: Further, we estimate parameter α,

$$\theta_0^{\alpha} = \frac{\ln\left|E\left\{e^{j\theta_0 x}\right\}\right|}{\ln\left|E\left\{e^{jx}\right\}\right|} = \frac{\ln\left|\hat{\varphi}(\theta_0;\alpha,\beta,\mu,\delta)\right|}{\ln\left|\hat{\varphi}(1;\alpha,\beta,\mu,\delta)\right|}, \tag{26}$$

$$\hat{\alpha} = \log_{\theta_0}\left(\frac{\ln\left|\hat{\varphi}(\theta_0;\alpha,\beta,\mu,\delta)\right|}{\ln\left|\hat{\varphi}(1;\alpha,\beta,\mu,\delta)\right|}\right), \tag{27}$$

where $\hat{\varphi}(\theta_0;\alpha,\beta,\mu,\delta) = \frac{1}{N}\left|\sum_{i=1}^{N} e^{j\theta_0 x_i}\right|$.

Step 4: Parameter μ is estimated by complex domain of the cumulant generating function of fLsm,

$$\ln\varphi(\theta_0;\alpha,\beta,\mu,\delta) = \delta|\theta|^{\alpha} + j\left[\delta|\theta|^{\alpha}\beta\frac{\theta}{|\theta|}\tan\left(\frac{\pi\alpha}{2}\right) + \mu\theta\right], \tag{28}$$

$$\hat{\mu} = \frac{Im\left\{\theta_0^{\hat{\alpha}}\ln\left|\hat{\varphi}(1;\alpha,\beta,\mu,\delta)\right| - \ln\left|\hat{\varphi}(\theta_0;\alpha,\beta,\mu,\delta)\right|\right\}}{\theta_0^{\alpha} - \theta_0}, \tag{29}$$

Step 5: As we know, the fLsm model drive function is symmetric $\hat{\beta} = 0$.

5. Wind Speed Forecasting

We used the average daily wind speed data from the 2011 actual historical wind speed of Inner Mongolia. The historical wind speed waveform is shown in Figure 2. When the wind speed is too high, it will seriously affect the power grid, so we focused on accurately predicting the time period when the wind speed is high. It can be seen from Figure 2 that the wind speed data began to fluctuate greatly from the 100th day, which was harmful to the power grid, so we chose to start from the 100th forecast. In terms of selecting the prediction steps, the small-data method of the second part was used to calculate the maximum prediction steps. The calculation results are shown in Table 1. The maximum forecast steps were 43 days, we could set the forecast time period from the 100th day to the 140th day. Before using the fLsm iterative forecasting model, we needed to determine whether the wind speed sequence was LRD. Through parameter estimation, we could get the value of H and α (Table 2), satisfying $\alpha H > 1$. Finally, the fLsm iterative forecasting model was used to forecast the wind speed sequence, and the forecast result is shown in Figure 3. The specific method flow is shown in Figure 4.

Table 1. 2011 Small-data method parameters.

Parameter Name	Parameter Value
Average period	12
Embedding Dim	5
Time delay	2
Lyapunov exponent	0.0238
Max. prediction steps	43

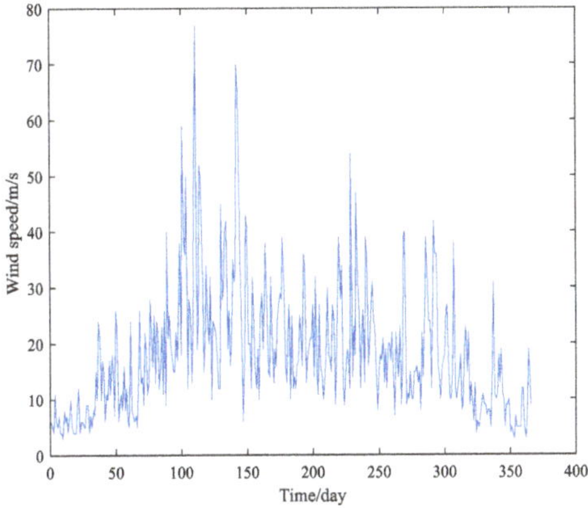

Figure 2. 2011 wind speed waveform.

Table 2. Parameters and errors of the three weighting methods.

Name	Unweighted	5 Weighted	10 Weighted
Max error percentage	3.7319	0.4425	0.1419
H	0.7595	0.7595	0.7595
α	1.7959	1.8280	1.6305
var	173.7598	4344	17376

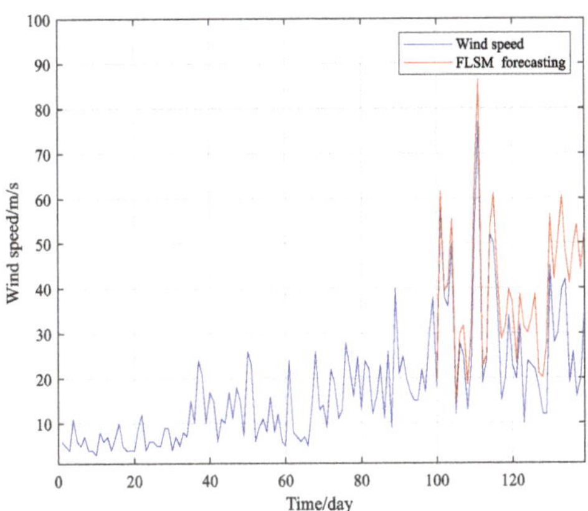

Figure 3. Unweighted wind speed predictions.

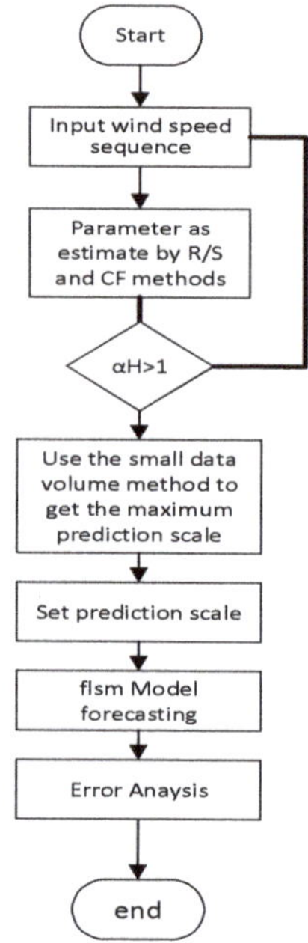

Figure 4. Forecasting process.

As can be seen from Figure 3, when the prediction steps exceeded nine steps, the prediction error gradually increased, and the prediction data was often larger than the actual data. However, by calculating the maximum prediction steps of 43, its effective prediction steps were much less than the maximum prediction steps.

As fLsm is an infinite variance process and the variance of the wind speed data is not large, if the historical wind speed data is used to estimate the parameters of the fLsm iterative prediction model, a large error will occur. In this section, we used a method of weighting the wind speed data to increase the variance of the data, thereby reducing the error in parameter estimation.

It can be seen from Figures 5 and 6 that the prediction effect of the wind speed weighted data had been significantly improved. Generally speaking, increasing the variance will cause the tail parameter α to decrease. However, it can be seen from Table 2 that the α value of the five-times weighted wind speed data was larger than the α value of the unweighted wind speed data, which indicated that a larger error occurred when modeling the wind speed sequence using the fLsm iterative prediction model. Of course, $\alpha H > 1$ must be guaranteed when weighting the wind speed data.

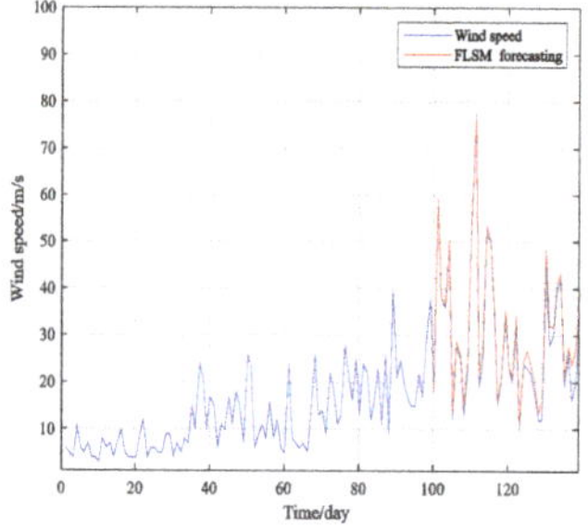

Figure 5. Five-time weighted wind speed predictions.

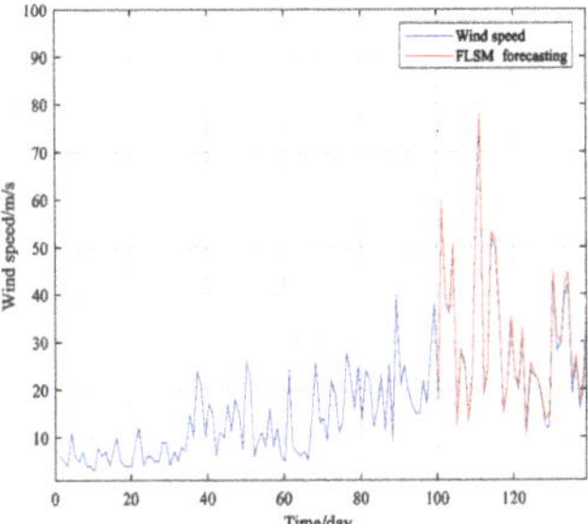

Figure 6. Ten-time weighted wind speed predictions.

In order to prove the extensiveness of the wind speed prediction model in this paper, we forecasted the wind speed data of Inner Mongolia in 2012. As can be seen from Figure 7, the wind speed data on the 70th day began to fluctuate. We then calculated the maximum number of prediction steps to 45 and set the prediction time period to the 70th to 115th days. After weighting the wind speed data 10 times, the fLsm iterative prediction model was used for prediction in Figure 8. In addition, the fLsm iterative prediction model was compared with the GA-BP neural network, which showed that the wind speed prediction model in this paper had better prediction accuracy.

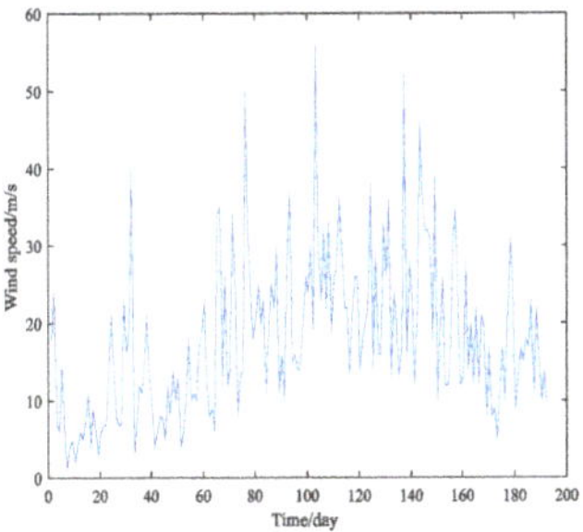

Figure 7. 2012 wind speed waveform.

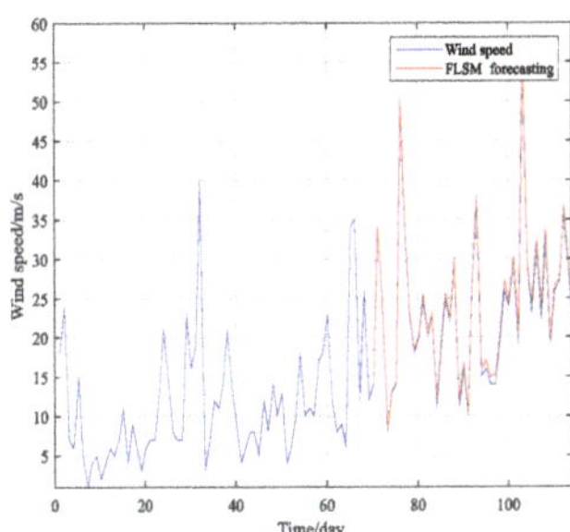

Figure 8. 2012 wind speed forecast results.

Table 3 lists the maximum and average percentage errors for the two prediction models. As can be seen from the table, the fLsm iterative prediction model had higher prediction accuracy. At the same time, it can be seen from Figures 9 and 10 that the GA-BP neural network had a poor prediction of the peak value, which will lead to the inability to prevent the impact of excessive wind speed on the grid.

Table 3. Errors of the two prediction models.

Years	Name	GA-BP Network	FLSM Forecasting
2011	Max error percentage	0.2706	0.1419
2011	Mean error percentage	0.0350	0.0304
2012	Max error percentage	0.2676	0.1022
2012	Mean error percentage	0.0378	0.0282

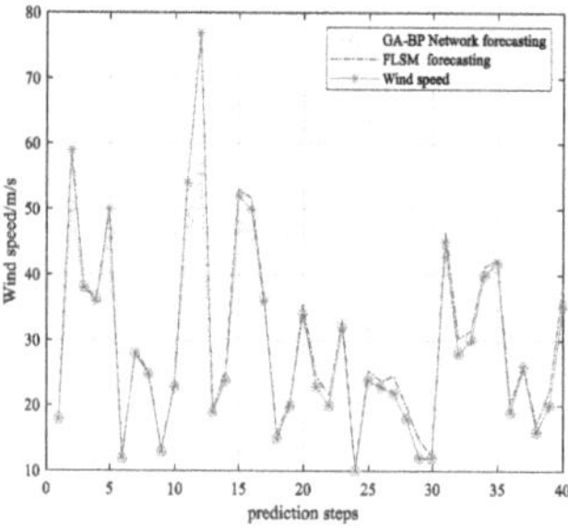

Figure 9. Comparison of prediction effects in 2011.

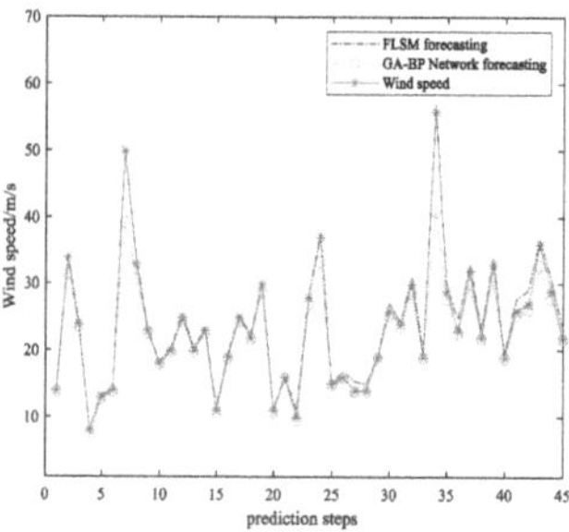

Figure 10. Comparison of prediction effects in 2012.

6. Relationship between Wind Speed and Wind Power

Taking a variable-pitch wind turbine with a single unit capacity of 600 kW as an example, the power characteristics are shown in Figure 11. The cut-in wind speed, cut-out wind speed, and rated wind speed were 3, 50, and 25 m/s, respectively. The raw data of wind power time series could be obtained from the original data of wind speed and power characteristic curve of the wind turbines.

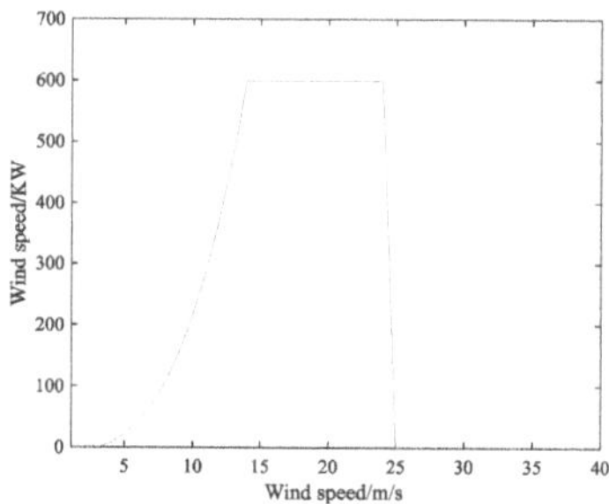

Figure 11. Power curve of a wind power generator.

When the wind speed was less than the cut-in wind speed and greater than the cut-out wind speed, the power generation was zero; when the wind speed was equal to the cut-in wind speed, the rated wind speed, and the cut-out wind speed, the power characteristic curve had a significant turning point. When the wind speed was greater than the rated wind speed and less than the cut-out wind speed when the wind speed was out, the generating power was a certain value. Only when the wind speed was greater than the cut-in wind speed and less than the rated wind speed did the generating power, and the wind speed approximate a linear relationship.

$$P(v) = \begin{cases} 0 & 0 \le v \le v_i \\ f_p(v) & v_i \le v \le v_r \\ P_r & v_r \le v \le v_c \\ 0 & v > v_c \end{cases}, \tag{30}$$

where $P(v)$ is the wind power, P_r is the rated power of the generator, v_i is the cut-in wind speed, v_c is the cut-out wind speed also known as the cut-off wind speed, v_r is the rated wind speed, and $f_p(v)$ is the output characteristic of the wind speed between v_i and v_r. Its characteristics can be linear functions, quadratic functions, or cubic functions.

7. Conclusions

(1) Wind speed prediction is of great significance to the stable operation and operating efficiency of the power system. At the same time, it improves the ability of wind farms to participate in market competition.

(2) The wind speed prediction method based on the maximum Lyapunov exponent and fLsm iterative prediction model was effective. Based on the historical wind speed sequence, this paper calculated the maximum prediction steps, weighted the wind speed data, and established an fLsm iterative prediction model. It can be seen from the MATLAB simulation curve that the model can better predict the wind speed and reflect the change of the sequence, which has certain guiding significance. It can be seen from Section 6 that after the conversion of the power characteristic curve, its regularity was partially destroyed, and the regularity of the obtained wind energy was even weaker, which led to a larger forecast error of wind power. Therefore, the wind speed needs to be predicted first, and then the amount of electricity can be calculated.

(3) In practice, wind speed has strong randomness, and some regions may not have LRD. The wind speed sequence of short-range dependent (SRD) has yet to be studied.

Author Contributions: Conceptualization, S.D. and C.C.; Data curation, Y.Y. and H.L.; Formal analysis, W.S. and C.C.; Funding acquisition, W.S.; Investigation, S.D., W.S. and C.C.; Methodology, S.D., W.S. and C.C.; Project administration, W.S. and C.C.; Resources, W.S. and C.C.; Visualization, W.S. and C.C.; Writing–Original draft, S.D.; Writing–Review & editing, W.S. and C.C. All authors have read and agreed to the published version of the manuscript.

Funding: This project was funded by the Natural Science Foundation of Shanghai (Grant No. 14ZR1418500).

Conflicts of Interest: The authors declare no conflict of interest.

References

1. Lei, Y.Z.; Wang, W.S.; Yin, Y.H.; Dai, H.Z. Analysis of wind power value to power system operation. *Power Syst. Technol.* **2002**, *5*, 62–66.
2. Shu-Yong, C.H.E.N.; Hui-Zhu, D.A.I.; Xiao-Min, B.A.I.; Xiao-Xin, Z.H.O.U. Reliability model of wind power plants and its application. *Proc. -Chin. Soc. Electr. Eng.* **2000**, *20*, 26–29.
3. Ya-Zhou, L.; Wei-Sheng, W.; Yong-Hua, Y. An optimization method for determining wind power penetration limit in power system under static security constraints. *Proc. -Chin. Soc. Electr. Eng.* **2001**, *21*, 25–28.
4. Lei, Y.Z.; Wang, W.S.; Yin, Y.H.; Dai, H.Z. Wind power penetration limit calculation based on chance constrained programming. *Proc. CSEE* **2002**, *22*, 32–35.

5. Alexiadis, M.C.; Dokopoulos, P.S.; Sahsamanoglou, H.S.; Manousaridis, I.M. Short term forecasting of wind speed and related electrical power. *Sol. Energy.* **1998**, *63*, 61–68. [CrossRef]
6. Bossanyi, E.A. Short-term wind prediction using Kalman filters. *Wind Eng.* **1985**, *9*, 1–8.
7. Kamal, L.; Jafri, Y.Z. Time series models to simulate and forecast hourly averaged wind speed in Wuetta, Pakistan. *Sol. Energy* **1997**, *61*, 23–32. [CrossRef]
8. Kariniotakis, G.N.; Stavrakakis, G.S.; Nogaret, E.F. Wind power forecasting using advanced neural network models. *IEEE Trans Energy Convers.* **1996**, *11*, 762–767. [CrossRef]
9. Bayraktar, E.; Poor, H.V.; Rao, R. Prediction and tracking of long-range dependent sequences. *Syst. Control Lett.* **2005**, *34*, 1083–1090. [CrossRef]
10. Gao, Y.; Villecco, F.; Li, M.; Song, W. Multi-Scale Permutation Entropy Based on Improved LMD and HMM for Rolling Bearing Diagnosis. *Entropy* **2017**, *19*, 176. [CrossRef]
11. Song, W.; Cattani, C.; Chi, C.H. Multifractional Brownian Motion and Quantum-Behaved Particle Swarm Optimization for Short Term Power Load Forecasting: An Integrated Approach. *Energy* **2020**, *194*, 116847. [CrossRef]
12. Wanqing, S.; Chen, X.; Cattani, C.; Zio, E. Multifractional Brownian Motion and Quantum-Behaved Partial Swarm Optimization for Bearing Degradation Forecasting. *Complexity* **2020**. [CrossRef]
13. Li, Y.; Song, W.; Wu, F.; Zio, E.; Zhang, Y. Spectral Kurtosis of Choi–Williams Distribution and Hidden Markov Model for Gearbox Fault Diagnosis. *Symmetry* **2020**, *12*, 285. [CrossRef]
14. Li, M. Fractal Time Series—A Tutorial Review. *Math. Probl. Eng.* **2010**. [CrossRef]
15. Liu, H.; Song, W.; Li, M.; Kudreyko, A.; Zio, E. Fractional Lévy stable motion: Finite difference iterative Forecasting model. *Chaos Solitons Fractals* **2020**, *133*, 109632. [CrossRef]
16. Rosenstein, M.T.; Collins, J.J.; De Luca, C.J. A practical method for calculating largest Lyapunov exponents from small data sets. *Phys. D: Nonlinear Phenom.* **1993**, *65*, 117–134. [CrossRef]
17. Kim, H.; Eykholt, R.; Salas, J.D. Nonlinear dynamics, delay time and embedding windows. *Phys. D: Nonlinear Phenom.* **1999**, *127*, 48–60. [CrossRef]
18. Tao, H.Y.C. Phase-space reconstruction technology of chaotic attractor based on c-c method. *J. Electron. Meas. Instrum.* **2012**, *26*, 425–430.
19. Weron, A.; Burnecki, K.; Mercik, S.; Weron, K. Complete description of all self-similar models driven by Levy stable noise. *Phys. Rev. E* **2005**. [CrossRef]
20. Black, F.; Scholes, M. The pricing of options and corporate liabilities. *J. Political Econ.* **1973**, *81*, 637–654. [CrossRef]
21. Jumarie, G. Merton's model of optimal portfolio in a Black-Scholes market driven by a fractional Brownian motion with short-range dependence. *Insur. Math. Econ.* **2005**, *37*, 585–598. [CrossRef]
22. Jumarie, G. On the representation of fractional Brownian motion as an integral with respect to (dt)a. *Appl. Math. Lett.* **2005**, *18*, 739–748. [CrossRef]
23. Qin, Y.; Xiang, S.; Chai, Y.; Chen, H. Macroscopic-microscopic attention in LSTM networks based on fusion features for gear remaining life prediction. *IEEE Trans. Ind. Electron.* **2020**. [CrossRef]
24. Wang, X.; Li, K.; Gao, P.; Meng, S. Research on parameter estimation methods for alpha stable noise in a laser gyroscope's random error. *Sensors* **2015**, *15*, 18550–18564. [CrossRef] [PubMed]
25. Ding, Z.; Sun, G.; Guo, M.; Jiang, X.; Li, B.; Liang, S.Y. Effect of phase transition on micro-grinding-induced residual stress. *J. Mater. Process. Technol.* **2020**, *281*, 116647. [CrossRef]
26. Zou, H.L.; Yu, Z.G.; Anh, V.; Ma, Y.L. From standard alpha-stable levy motions to horizontal visibility networks: Dependence of multifractal and Laplacian spectrum. *J. Stat. Mech. Theory Exp.* **2018**, 053403. [CrossRef]
27. Song, W.; Li, M.; Li, Y.; Cattani, C.; Chi, C.H. Fractional Brownian motion: Difference iterative forecasting models. *Chaos Solitons Fractals* **2019**, *123*, 347–355. [CrossRef]
28. Kogon, S.M.; Manolakis, D.G. Signal modeling with self-similar stable processes: The fractional levy stable motion model. *IEEE Trans. Signal Process.* **1996**, *44*, 1006–1010. [CrossRef]
29. Laskin, N.; Lambadaris, I.; Harmantzis, F.C.; Devetsikiotis, M. Fractional levy motion and its application to network traffic modeling. *Comput. Netw.* **2002**, *40*, 363–375. [CrossRef]
30. Karasaridis, A.; Hatzinakos, D. Network heavy traffic modeling using stable self-similar processes. *Ieee Trans. Commun.* **2001**, *49*, 1203–1214. [CrossRef]

31. Dai, W.; Heyde, C.C. It's formula with respect to fractional brownian motion and its application. *J. Appl. Math. Stoch. Anal.* **1996**, *9*. [CrossRef]
32. Wang, X.T.; Qiu, W.Y.; Ren, F.Y. Option pricing of fractional version of the black-scholes model with hurst exponent h being in (1/3,1/2). *Chaos Solitons Fractals* **2001**, *12*, 599–608. [CrossRef]

Article

On Two-Dimensional Fractional Chaotic Maps with Symmetries

Fatima Hadjabi [1,*], Adel Ouannas [2], Nabil Shawagfeh [1], Amina-Aicha Khennaoui [3] and Giuseppe Grassi [4]

1 Department of Mathematics, The University of Jordan, Amman 11942, Jordan; shawagnt@ju.edu.jo
2 Laboratory of Mathematics, Informatics and Systems (LAMIS), University of Larbi Tebessi, Tebessa 12002, Algeria; ouannas.adel@univ-tebessa.dz
3 Laboratory of Dynamical Systems and Control, University of Larbi Ben M'hidi, Oum El Bouaghi 4003, Algeria; Khennaoui.amina@univ-leb.dz
4 Dipartimento Ingegneria Innovazione, Universita del Salento, 73100 Lecce, Italy; giuseppe.grassi@unisalento.it
* Correspondence: fatimahdj3@gmail.com

Received: 8 March 2020; Accepted: 8 April 2020; Published: 6 May 2020

Abstract: In this paper, we propose two new two-dimensional chaotic maps with closed curve fixed points. The chaotic behavior of the two maps is analyzed by the 0–1 test, and explored numerically using Lyapunov exponents and bifurcation diagrams. It has been found that chaos exists in both fractional maps. In addition, result shows that the proposed fractional maps shows the property of coexisting attractors.

Keywords: discrete fractional systems; chaotic systems; closed curve fixed points; symmetry; 0–1 test; bifurcation diagram; Lyapunov exponents

1. Introduction

In the nineteenth century, fractional calculus had its origin in the generalization of integer order differentiation and integration to non-integer order (fractional order) ones [1–5]. The first treatise concerning fractional calculus was conducted by K.B Oldham and J. Spanier (1974) [6]. In the early stages of fractional calculus, the focus was on continuous time fractional dynamical systems [7]. A few decades later, interest shifted to discrete fractional calculus. Diaz and Olser (1974) gave the first definition of a discrete fractional operator. It was introduced by discretization of the continuous time fractional operator, and this led to representations by recurrence relations. The first definition of fractional order operators was pioneered by Miler and Ross [8]. Following these definitions, and in a series of papers, Atici and Eloe [1,9–11] defined and proved many properties of these operators. Balanu et al. provided significant contributions for the subject concerning stability, and chaos of some discrete fractional systems [4,12–14]. Discrete fractional calculus is a sub-discipline of dynamical system that has been of significant importance since its first emergence. It first appeared in relation to real-world phenomena, and in a broad range of disciplines such as biology, economics, physics, engineering [15], nonlinear optics [16], finance [17], demography [18], medicine [19], and so on. It has become an interesting field of research in the last decades.

The concept of chaos in dynamical systems has attracted researchers due to the unpredictability of the dynamic behaviors of dynamical systems. This means that two systems may start very close, but they move very far for apart. In fact, Li and York [20] were the first researchers who gave a mathematical definition of chaos. More studies and investigations for a variety of chaotic maps have been presented in [21]. In addition, many considerable results have been proposed in [14,21–24]. Those maps possess sensitivity to initial conditions and random-like trajectories. One of the characteristics of chaos is the existence of at least one positive Lyapunov exponent. Another efficient method for detecting chaos is the 0–1 test. That is, if the output is zero, this means that the dynamic is regular; it is chaotic if the test yields one [25]. The fact that research in discrete fractional chaotic systems is still in development [26–34], and very few difference equations have been considered, was the motivation of our work.

The aim of this paper is to study two new chaotic maps with specific types of fixed points by the application of a new test approach (0–1 test) in order to find out whether or not these systems are chaotic. The dynamic behaviors of the considered chaotic systems are investigated numerically using bifurcation diagrams and Lyapunov exponents. These systems possess an interesting property: symmetry. The rest of the paper is organized as follows. In Section 2, we present certain essential definitions and results from difference fractional calculus. Then, chaotic dynamics of the two systems will be discussed in Section 3. Section 4 is about the chaotic analysis based on the 0–1 test approach. Finally, we conclude with a summary.

2. Preliminaries

We give in this section some basic definitions, and one major theorem from discrete fractional calculus. Let $X : \mathbb{N}_a \to \mathbb{R}$, with $\mathbb{N}_a = \{a, a+1, a+2, \dots\}$, the general n^{th} order difference can be written as:

$$\Delta^n x(t) = \Delta^{n-1} X(t+1) - \Delta^{n-1} X(t) = \sum_{k=0}^{n} C_k^n (-1)^k X(t+n-k). \tag{1}$$

Prolonging this concept for fractional-order difference, the fractional sum of order v, using the fact that $t^{(v)}$ is defined by $t^{(v)} = \frac{\Gamma(t+1)}{\Gamma(t+1-v)}$, is given in the following definition.

Definition 1 ([35]). *The vth fractional sum of X can be defined as:*

$$\Delta_a^{-v} X(t) = \frac{1}{\Gamma(v)} \sum_{s=a}^{t-v} (t - \sigma(s))^{(v-1)} X(s), \tag{2}$$

where $v \notin \mathbb{N}$, with $v > 0$ is a fractional order, $t \in \mathbb{N}_{a+n-v}$, $n = [v] + 1; \sigma(t) = t + 1$.

Definition 2 ([35]). *The v-Caputo-like difference, denoted by ${}^C\Delta_a^v X(t)$, of $X(t)$ is defined as:*

$${}^C\Delta_a^v X(t) = \Delta_a^{-(n-v)} \Delta^n X(t) = \frac{1}{\Gamma(n-v)} \sum_{s=a}^{t-(n-v)} (t - \sigma(s))^{(n-v-1)} \Delta_a^n X(s). \tag{3}$$

Theorem 1 ([35]). *The corresponding discrete integral equation of the delta fractional difference equation:*

$$\begin{cases} {}^C\Delta_a^v u(t) & = f(t+v-1, u(t+v-1)) \\ {}^C\Delta^k u(a) & = u_k, \quad n = [v]+1, \quad k = 0, \dots, n-1, \end{cases} \tag{4}$$

is given by:

$$u(t) = u_0(t) + \frac{1}{\Gamma(v)} \sum_{s=a+n-v}^{t-v} (t - \sigma(s))^{(v-1)} f(s+v-1, u(s+v-1)), \ t \in \mathbb{N}_{a+n}, \tag{5}$$

where:

$$u_0(t) = \sum_{k=0}^{n-1} \frac{(t-a)^k}{k} \Delta^k u(a). \tag{6}$$

3. Fractional–Order Maps with Closed Curve Fixed Points

In this section, we study the chaotic behavior of two new two-dimensional fractional systems with closed curve fixed points. In [36], a new class of two-dimensional chaotic maps with different types of closed curve fixed points are constructed. Examples of these systems are given by:

$$\begin{cases} x_{k+1} &= x_k + \alpha f(x_k, y_k), \\ y_{k+1} &= y_k + f(x_k, y_k)(\beta x_k + \gamma y_k + \delta x_k^2 + \epsilon y_k^2 + \varepsilon x_k y_k + \theta), \end{cases} \tag{7}$$

where x_k, y_k are system states, $\{\alpha, \beta, \gamma, \delta, \epsilon, \varepsilon, \theta\}$ are system parameters and $f(x, y)$ is a nonlinear function which can be written as

$$f(x,y) = (\frac{x}{m})^p + (\frac{y}{n})^p - r^2, \tag{8}$$

where $m > 0, n > 0, r > 0$, and p is an integer order such that $p \leq 2$. The authors showed that the fixed points of these systems are nonhyperbolic for different values of m, n, p, and r, and they presented the phase-basin portrait of the chaotic attractors. Also, they explored them numerically by Lyapunov exponents and Kaplan–Yorke dimension.

3.1. Fractional–Order Map with Square-Shaped Fixed Points

Here we consider the system (7) when $m = 1, n = 1, p = 12$, and $r = 1$. In this case, the map is described by the following equation:

$$\begin{cases} x_{k+1} &= x_k - \alpha(x_k^{12} + y_k^{12} - 1), \\ y_{k+1} &= y_k + \beta x_k y_k (x_k^{12} + y_k^{12} - 1), \end{cases} \tag{9}$$

where α and β are bifurcation parameters. Based on the above equation, we construct a new fractional-order map by introducing the υ-Caputo like difference operator as follows:

$$\begin{cases} {}^C\Delta_a^{\upsilon_1} x(t) &= -\alpha(x^{12}(t-1+\upsilon_1) + y^{12}(t-1+\upsilon_1) - 1), \\ {}^C\Delta_a^{\upsilon_2} y(t) &= \beta x(t-1+\upsilon_2) y(t-1+\upsilon_2)(x^{12}(t-1+\upsilon_2) + y^{12}(t-1+\upsilon_2) - 1), \end{cases} \tag{10}$$

where $\upsilon = (\upsilon_1, \upsilon_2)$ is the fractional order. The vector field $F(x, y)$ associated to the system (10) possesses a symmetry with respect to the x-axis. That is, if we let $G(x, y) = (x, -y)$, then $F(G(x, y)) = G(F(x, y))$. This property will be reflected in the fixed points and attractors of system (10). The fixed points of the new fractional-order system (10) are obtained by taking ${}^C\Delta_a^{\upsilon_1} x = {}^C\Delta_a^{\upsilon_2} y = 0$, as follows:

$$\begin{cases} -\alpha(x^{12} + y^{12} - 1) &= 0, \\ \beta xy(x^{12} + y^{12} - 1) &= 0. \end{cases} \tag{11}$$

From Equation (11) we must have

$$x^{12} + y^{12} - 1 = 0. \tag{12}$$

Consequently, the system (10) has square-shaped fixed points. By choosing suitable system parameters the fractional-order map (10), we can generate hidden attractors with square-shaped fixed points.

In order to further investigate the dynamical behavior of the new fractional-order map (10), we need to determine the corresponding numerical formula. Using Theorem 1, the equivalent discrete integral equation of the fractional-order map (10) can be written as:

$$\begin{cases} x(t) = x(a) - \frac{\alpha}{\Gamma(v_1)} \sum_{s=a+1-v_1}^{t-v_1} (t-s-1)^{(v_1-1)} (x^{12}(s-1+v_1) + y^{12}(s-1+v_1) - 1), \\ y(t) = y(a) + \frac{\beta}{\Gamma(v_2)} \sum_{s=a+1-v2}^{t-v_2} (t-s-1)^{(v_2-1)} \\ \qquad x(s-1+v_2)y(s-1+v_2)(x^{12}(s-1+v_2) + y^{12}(s-1+v_2) - 1), \end{cases} \tag{13}$$

in which $t \in \mathbb{N}_{a+1-v}$ and a is the starting point. Replacing the discrete kernel function $\frac{(t-\sigma(s))^{(v-1)}}{\Gamma(v)}$ by $\frac{\Gamma(t-s)}{\Gamma(v)\Gamma(t-s-v+1)}$ and assuming that $a = 0$, the above equation is converted to:

$$\begin{cases} x(n) = x(0) - \frac{\alpha}{\Gamma(v_1)} \sum_{j=1}^{n} \frac{\Gamma(n-j+v_1)}{\Gamma(n-j+1)} (x^{12}(j-1) + y^{12}(j-1) - 1), \\ y(n) = y(0) + \frac{\beta}{\Gamma(v_2)} \sum_{j=1}^{n} \frac{\Gamma(n-j+v_2)}{\Gamma(n-j+1)} \left(x(j-1)y(j-1)\left(x^{12}(j-1) + y^{12}(j-1)\right)\right). \end{cases} \tag{14}$$

where $x(0)$ and $y(0)$ are the initial condition. This numerical formula will allow us to examine the sensitivity of the fractional-order map throughout the next subsection.

Bifurcation Diagrams and Largest Lyapunov Exponents

In this section, we will study the dynamical behavior of the novel fractional-order map (10) for some specific parameters and initial values. Figure 1 shows the chaotic attractor along with its bifurcation plots with respect to α and β, and the time evolution of states x for $v = 1$. The chaotic attractor of the fractional-order map (10) under system parameters $\alpha = 0.2, \beta = 2.4$ and initial value $(x(0), y(0)) = (0.26, -0.14)$ is shown in Figure 1a, where the black points represent the square-shaped fixed points. Figure 1b gives the time evolutions of the variables x and y which illustrates that the system (10) has two symmetrical states. Figure 1c,d shows the bifurcation plots produced when fixing one of the parameters and varying the second. These bifurcation diagrams are constructed for 50 initial points. We see that the maximum chaotic range is observed for $\alpha \approx 0.35$ and $\beta \approx 2.455$. Also, we notice that the range of β for which we obtain a chaotic behavior is very short. Hence, it is more interesting to visualize the effect of α on the map's dynamics.

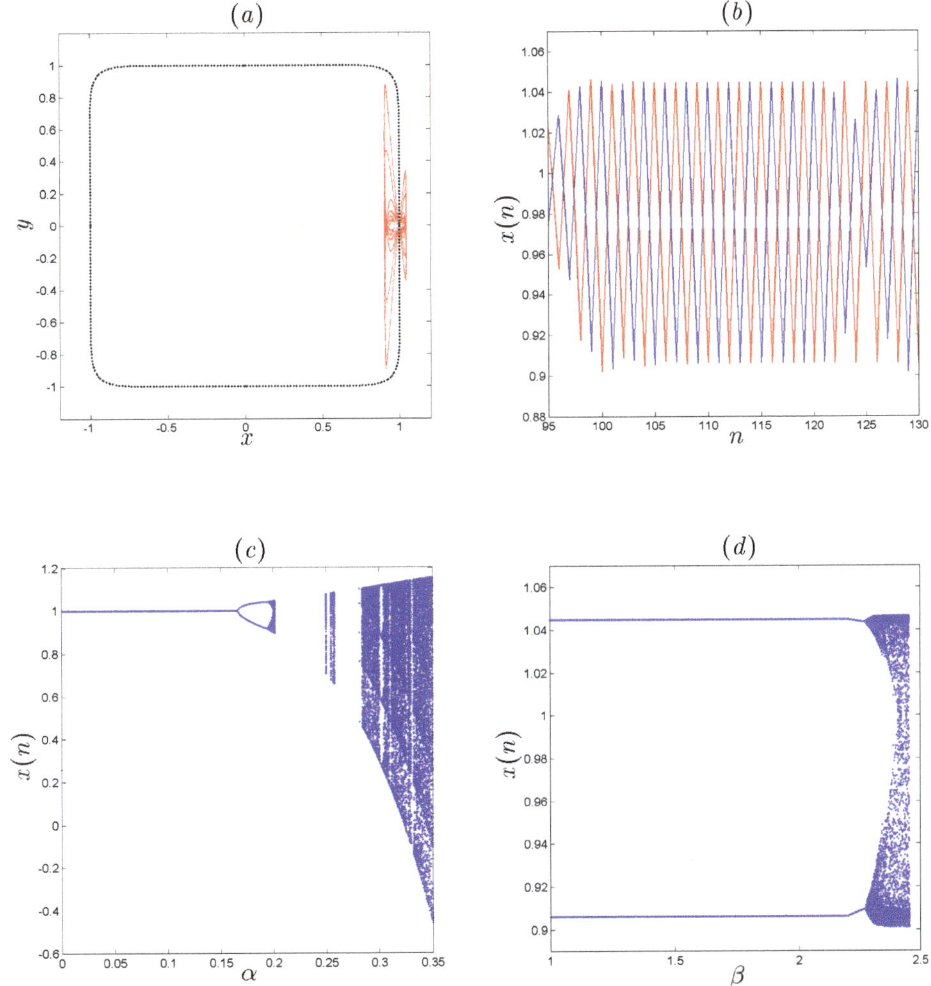

Figure 1. Numerical analysis of the fractional-order map (10) for $\nu = 1$. (**a**) Square-shaped fixed point of the fractional-order map are illustrated with black dots and the hidden chaotic attractor in red. (**b**) Evolution states of the fractional-order map (10) of x and y. (**c**) Bifurcation diagram of the fractional-order map (10) versus α for $\beta = 2.4$. (**d**) Bifurcation diagram of the fractional-order map (10) versus β for $\alpha = 0.2$.

To investigate the sensitivity of the fractional-order map (10) with respect to the bifurcation parameter α, we fix $\beta = 2.4$ and we vary α in the range $[0, 0.35]$. The bifurcation diagrams and largest Lyapunov exponents under the fractional order $v_1 = v_2 = 0.95$ and $v_1 = v_2 = 0.9$ are given in Figure 2. By comparing Figure 2a,b with Figure 2c,d, we find that the interval where chaos exists shrinks as v decreases. In particular, when the value of $v = (v_1, v_2)$ changes from 0.95 to 0.9, the areas of chaotic motion shrinks from $[1.022, 1.116] \cup [1.164, 1.216]$ to $[0.2405, 0.2429] \cup [0.3126, 0.3423]$. Also, the transient states region increases.

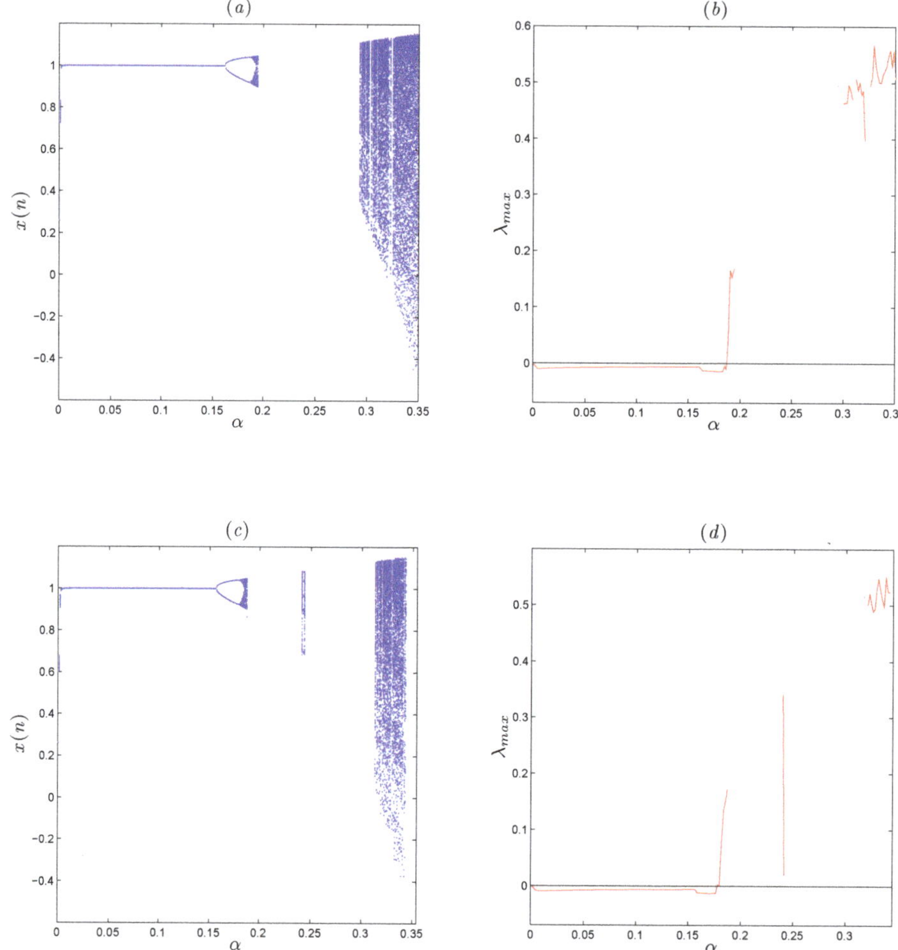

Figure 2. (**a**) Bifurcation diagram of the fractional-order map (10) versus α for $v_1 = v_2 = 0.95$. (**b**) Largest Lyapunov exponents corresponding to (**a**). (**c**) Bifurcation diagram of the fractional-order map (10) versus α for $v_1 = v_2 = 0.9$. (**d**) Largest Lyapunov exponents corresponding to (**c**).

Now, reset the parameter $\beta = 2.2$ and keep the parameter α as $\alpha = 0.2$, Figure 3 shows the bifurcation diagram of the fractional-order map (10) with the decreasing of v_2 where $v_1 = 1$. When v_2 decreases from 1, the fractional-order map (10) presents a periodic state until it enters into chaos at 0.9035. It is noticed that the periodic state of the integer order map becomes chaotic as v_2 decreases. Another interesting dynamic is observed if we increase the value of α to 0.23 and decrease the value of β to 0.5. In this case, two chaotic attractors are observed in the fractional-order map with respect to the fractional orders $v_1 = 1$ and $v_2 = 0.01$, as shown in Figure 4, where the red color attractor is yielded from initial value $(0.1, 0.3)$ and blue color attractor is yielded from initial value $(0.1, -0.3)$. Therefore, we conclude that the fractional order improves the complexity of the chaotic motion of the original system (7) with square-shaped fixed points.

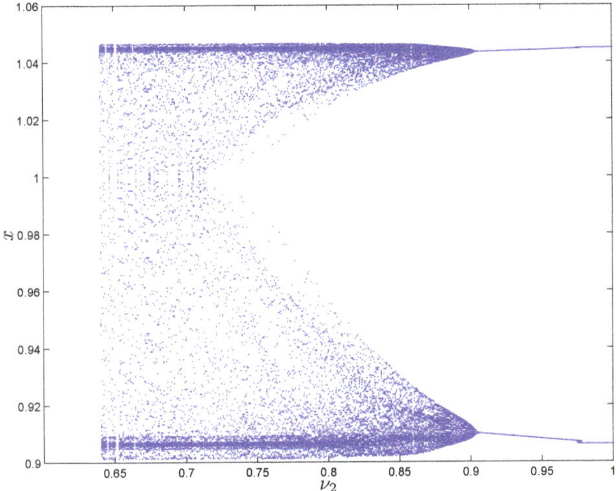

Figure 3. Bifurcation diagram for system (10) in the $x - v_2$ plane for $\alpha = 0.2$ and $\beta = 2.2$.

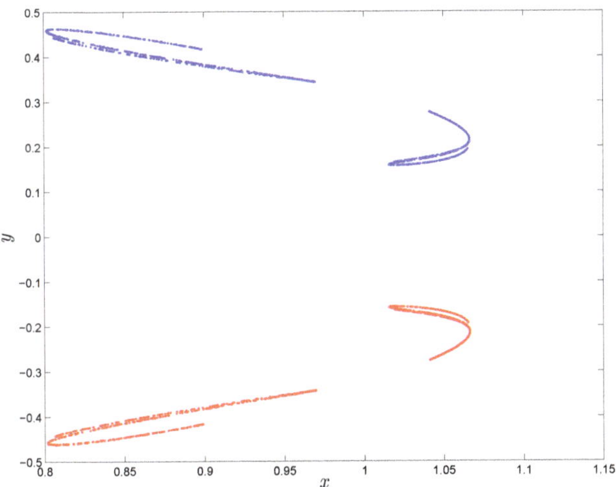

Figure 4. Coexisting chaotic attractor of the fractional-order map (16) for $\alpha = 0.2$ and $\beta = 2.2$.

3.2. A New Fractional Map with Rectangle-Shaped Fixed Points

Now, we consider system (7) when $m = 5$, $n = 1$, $p = 12$, $r = 1$. In this case we have the following system:

$$\begin{cases} x_{k+1} &= x_k + ((\frac{x_k}{5})^{12} + y_k^{12} - 1), \\ y_{k+1} &= y_k + \alpha y_k((\frac{x_k}{5})^{12} + y_k^{12} - 1), \end{cases} \tag{15}$$

where x and y are the system variables. Using the same argument as with the previous system, we obtain the following fractional-order map:

$$\begin{cases} {}^C\Delta_a^{v_1} x(t) = \left(\frac{x(t-1+v_1)}{5}\right)^{12} + y^{12}(t-1+v_1) - 1, \\ {}^C\Delta_a^{v_2} y(t) = \alpha y(t-1+v_2)((\frac{x(t-1+v_2)}{5})^{12} + y^{12}(t-1+v_2) - 1), \end{cases} \tag{16}$$

where $v \in (0,1)$ is the fractional order. The fixed points of the new fractional-order map (16) are obtained by solving the following system of equation:

$$\begin{cases} \left(\frac{x}{5}\right)^{12} + y^{12} - 1 & = 0, \\ \alpha y((\frac{x}{5})^{12} + y^{12} - 1) & = 0. \end{cases} \tag{17}$$

Thus, we must have

$$(\frac{x}{5})^{12} + y^{12} - 1 = 0. \tag{18}$$

Therefore, the novel fractional-order map (16) has rectangle-shaped fixed points.

To study the dynamic behavior of the novel fractional-order map (16), we define the numerical formula as:

$$\begin{cases} x(n) = x(0) + \frac{1}{\Gamma(v_1)} \sum_{j=1}^{n} \frac{\Gamma(n-j+v_1)}{\Gamma(n-j+1)} \left(\left(\frac{x(j-1)}{5}\right)^{12} + \left(\frac{y(j-1)}{12}\right)^{12} - 1 \right), \\ y(n) = y(0) + \frac{\alpha}{\Gamma(v_2)} \sum_{j=1}^{n} \frac{\Gamma(n-j+v_2)}{\Gamma(n-j+1)} \left(y(j-1) \left(\left(\frac{x(j-1)}{5}\right)^{12} + y^{12}(j-1) - 1 \right) \right). \end{cases} \tag{19}$$

where $x(0)$ and $y(0)$ are the initial conditions.

Bifurcation and Largest Lyapunov Exponents

In this section we investigate the dynamic behavior of the new fractional-order map (16) using numerical simulations. When $\alpha = 2.2$ the fractional-order map (16) shows the hidden chaotic attractor as depicted in Figure 5. It can be shown that it also has a symmetry with respect to the x-axis. Figure 6 shows the bifurcation diagrams and largest Lyapunov exponents of the fractional-order map (16). According to Figure 6, it is very clear that the dynamic behavior of system (16) goes from periodic state to chaotic state with the increase of the control parameter α. Obviously, as the fractional order $v = (v_1, v_2)$ decreases from 0.998 to 0.995, the bifurcation diagram gradually shrinks to the left.

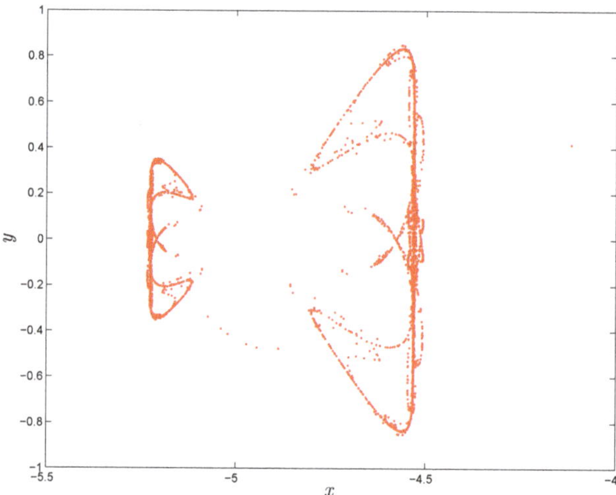

Figure 5. Chaotic attractor of the fractional-order map (16) for $\alpha = 2.2$.

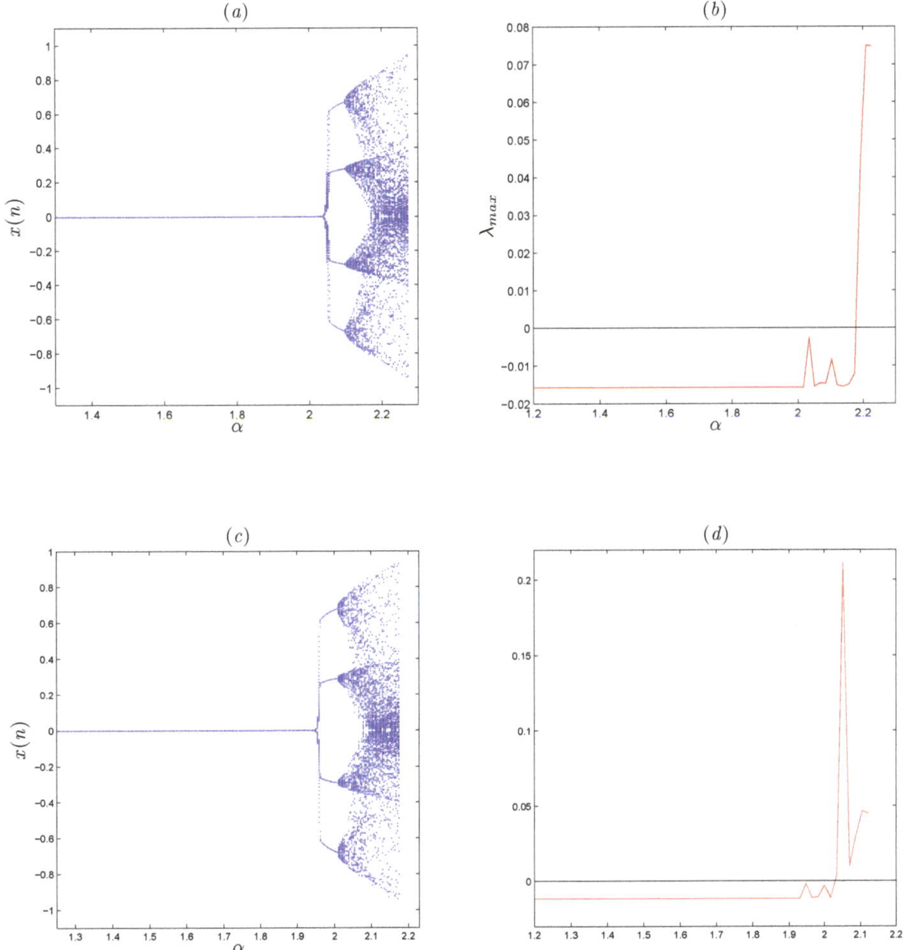

Figure 6. (**a**) Bifurcation diagram of the fractional-order map (16) versus α for $v_1 = v_2 = 0.988$. (**b**) Largest Lyapunov exponents corresponding to (**a**). (**c**) Bifurcation diagram of the fractional-order map (16) versus α for $v_1 = v_2 = 0.995$. (**d**) Largest Lyapunov exponents corresponding to (**c**).

3.3. 0–1 Test

In order to further confirm the influence of the fractional order $v = (v_1, v_2)$ on the properties of the fractional-order maps, we apply a new approach proposed by Gottwald and Melbourne called the 0–1 test method [25]. Unlike other methods, the test does not involve phase space reconstruction, and we can applied it directly to the series data $x(n)$. The output is either 0 or 1 depending on whether the nature of the system is regular or chaotic. Naturally, we do not obtain the values 1 and 0 exactly, so the test remains valid when K is close enough to these values.

Here, we applied the 0–1 test method directly to the state $x(n)$ that was obtained from the numerical formulas (14) and (19), respectively. For an arbitrary constant c in $(0, \pi)$, we calculate the translation variables

$$p_c(n) = \sum_{j=1}^{n} x(j) \cos(jc), \quad q_c(n) = \sum_{j=1}^{n} x(j) \sin(jc), \quad n = 1, 2, .., N. \tag{20}$$

The dynamics of the translation components (p_c, q_c) provide a visual test. Basically, if the dynamic is regular then the behavior of trajectories in the $(p_c - q_c)$ plane is bounded, whereas if the dynamic is chaotic then the $(p_c - q_c)$ trajectories depict Brownian like behavior. In order to examine the diffusive behavior of p_c and q_c, we define the mean square displacement as:

$$M_c(n) = \lim_{N \to \infty} \frac{1}{N} \sum_{j=1}^{N} \left((p_c(j+n) - p_c(j))^2 + (q_c(j+n) - q_c(j))^2 \right), \quad n \leq \frac{N}{10}, \tag{21}$$

with the asymptotic growth rate of

$$K_c = \lim_{n \to \infty} \frac{\log M_c(n)}{\log n}. \tag{22}$$

In practice, the final result K can be determined numerically by computing the median of K_c. When $K \simeq 1$, the behavior is classified as chaotic and when K is close to 0 the behavior is regular.

In Figure 7, we present the 0–1 test of the fractional-order map (10), when $v = 0.95$ and $\alpha = 0.2$. It is clear that the output K converges to the values 1, which implies that system (10) is chaotic. Moreover, when $0.98 \leq v < 0.998$ the fractional-order map (10) has a transient state as shown in Figure 8. As it can be seen that the state first converges into a bounded attractor and then it becomes divergent.

Similarly, for the second fractional-order map (16), we use Equation (19). The 0–1 test method is applied for $v = 0.998$, and $v = 0.995$, and the resulting diagrams are plotted in Figure 9. It shows that the asymptotic growth rate K approaches 1 as α increases, which confirms very well the results in Figure 6.

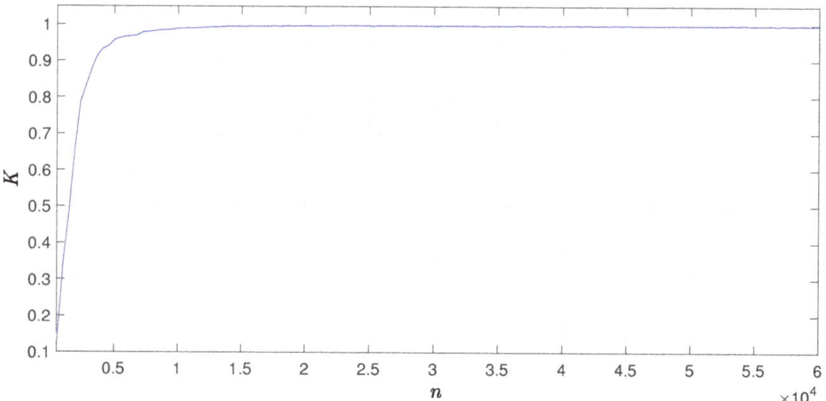

Figure 7. The 0–1 test method of the fractional-order map (10).

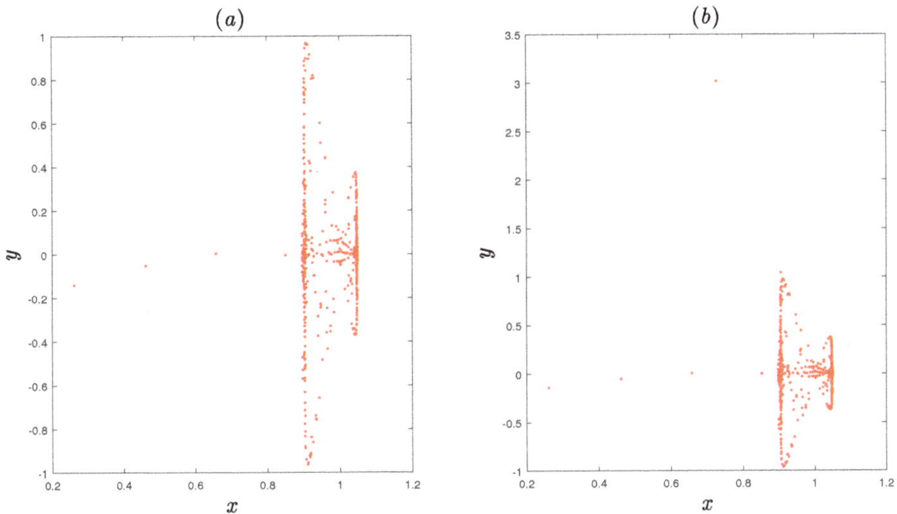

Figure 8. Transient state of the fractional-order map for $v = 0.98$ (**a**) when $n = 570$, (**b**) when $n = 578$.

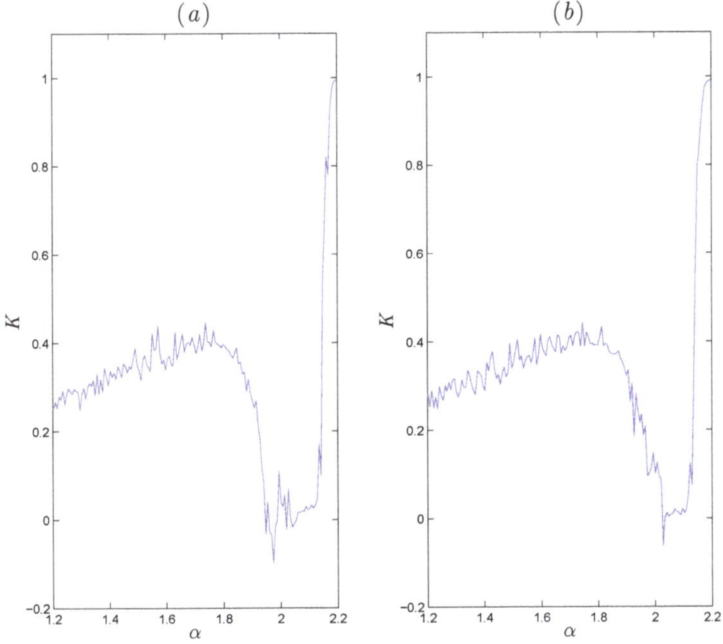

Figure 9. The 0–1 test of the fractional-order map (16). (**a**) The asymptotic growth rate versus α for $v_1 = v_2 = 0.998$. (**b**) The asymptotic growth rate versus α for $v_1 = v_2 = 0.995$.

4. Conclusions

In this article, we investigated the chaotic behavior of new two-dimensional fractional chaotic maps with closed curve fixed points. Numerical methods, including computations of Lyapunov exponents, bifurcation diagrams, and the 0–1 test have been used to illustrate the complex dynamics of these systems. Through this paper, we have shown that chaos exists in these fractional-order maps and that the type and range of chaotic behavior is dependent on the fractional order. Results shows that the fractional-order maps are more complex then their integer order counterparts. Also, numerical experiments have shown that the system exhibits the property of coexisting attractors. Future studies will explore more dynamic behaviors of those systems, such as, control and synchronization.

Author Contributions: Conceptualization, A.O., F.H.; N.S., and A.-A.K.; Data curation, A.-A.K. and F.H.; A.O. Investigation, F.H., A.O., and N.S.; Methodology, A.O., F.H., Supervision, N.S., A.O., and G.G.; Validation, A.O. and N.S. All authors have read and agreed to the published version of the manuscript.

Acknowledgments: The author Adel Ouannas was supported by the Directorate General for Scientific Research and Technological Development of Algeria.

Conflicts of Interest: The authors declare no conflict of interest.

References

1. Atıcı, F.M.; Eloe, P.W. Discrete fractional calculus with the nabla operator. *Electron. J. Qual. Theory Differ. Equ.* **2009**, 1–12. [CrossRef]
2. Abdeljawad, T.; Baleanu, D.; Jarad, F.; Agarwal, R.P. Fractional sums and differences with binomial coefficients. *Discret. Dyn. Nat. Soc.* **2013**, 1–6. [CrossRef]
3. Abdeljawad, T. On Riemann and Caputo fractional differences. *Comput. Math. Appl.* **2011**, *62*, 1602–1611. [CrossRef]
4. Baleanu, D.; Wu, G.C.; Bai, Y.R.; Chen, F.L. Stability analysis of Caputo–like discrete fractional systems. *Commun. Nonlinear Sci. Numer. Simul.* **2017**, *48*, 520–530. [CrossRef]
5. Goodrich, C.; Peterson, A.C. *Discrete Fractional Calculus*; Springer: Berlin, Germany, 2015; ISBN 978-3-319-79809-7.
6. Ross, B. The development of fractional calculus 1695–1900. *Hist. Math.* **1977**, *4*, 75–89. [CrossRef]
7. Kilbas, A.A.A.; Srivastava, H.M.; Trujillo, J.J. *Theory and Applications of Fractional Differential Equations*; Elsevier Science Limited: New York, NY, USA, 2006; Volume 204, ISBN 9780444518323.
8. Miller, K.S.; Ross, B. Fractional Difference Calculus. In *Proceedings of the International Symposium on Univalent Functions, Fractional Calculus and Their Applications, Nihon University, Koriyama, Japan, May 1988*; Ellis Horwood Ser. Math. Appl.; Horwood: Chichester, UK, 1989; pp. 139–152.
9. Atici, F.M.; Eloe, P.W. A transform method in discrete fractional calculus. *Int. J. Differ. Equ.* **2007**, *2*, 165–176.
10. Atıcı, F.M.; Eloe, P.W. Two-point boundary value problems for finite fractional difference equations. *J. Differ. Equ. App.* **2011**, *17*, 445–456. [CrossRef]
11. Atici, F.; Eloe, P. Initial value problems in discrete fractional calculus. *Am. Math. Soc.* **2009**, *137*, 981–989. [CrossRef]
12. Wu, G.C.; Baleanu, D. Discrete fractional logistic map and its chaos. *Nonlinear Dyn.* **2014**, *75*, 283–287. [CrossRef]
13. Wu, G.C.; Baleanu, D.; Zeng, S.D. Discrete chaos in fractional sine and standard maps. *Phys. Lett. A* **2014**, *378*, 484–487. [CrossRef]
14. Wu, G.C.; Baleanu, D.; Huang, L.L. Novel Mittag-Leffler stability of linear fractional delay difference equations with impulse. *Appl. Math. Lett.* **2018**, *82*, 71–78. [CrossRef]
15. Strogatz, S.H. *Nonlinear Dynamics and Chaos: With Applications to Physics, Biology, Chemistry, and Engineering*, 2nd ed.; Westview Press: Boulder, CO, USA, 2015; ISBN 978-0813349107.
16. Uchida, A. *Optical Communication with Chaotic Lasers: Applications of Nonlinear Dynamics and Synchronization*, 1st ed.; John Wiley & Sons: Weinheim, Germany, 2012; ISBN 9783527408696.
17. Guegan, D. Chaos in economics and finance. *Annu. Rev. Control.* **2009**, *33*, 89–93. [CrossRef]
18. Ugarcovici, I.; Weiss, H. Chaotic dynamics of a nonlinear density dependent population model. *Nonlinearity* **2004**, *17*, 1689–1711. [CrossRef]
19. Sengul, S. Discrete Fractional Calculus and Its Applications to Tumor Growth. Master's Theses, Western Kentucky University, Bowling Green, KY, USA, 1 May 2010; p. 161, ISBN 699811846.
20. Akin, E.; Kolyada, S. Li-Yorke sensitivity. *Nonlinearity* **2003**, *16*, 1421–1433. [CrossRef]
21. Liu, Y. Chaotic synchronization between linearly coupled discrete fractional Hénon maps. *Indian J. Phys.* **2010**, *90*, 313–317. [CrossRef]

22. Lozi, R. Un attracteur étrange du type attracteur de Hénon. *Le Journal de Physique Colloques* **1978**, *39*, C5–C9. [CrossRef]
23. Shukla, M.K.; Sharma, B.B. Investigation of chaos in fractional order generalized hyperchaotic Henon map. *Aeu-Int. J. Electron. C* **2017**, *78*, 265–273. [CrossRef]
24. Li, C.; Chen, Y.; Kurths, J. Fractional calculus and its applications. *Philos. Tr. Soc. A* **2013**, *371*, 20130037. [CrossRef]
25. Gottwald, G.A.; Melbourne, I. The 0-1 test for chaos: A review. In *Chaos Detection and Predictability*; Springer: Berlin/Heidelberg, Germany, 2016; pp. 221–247. [CrossRef]
26. Ouannas, A.; Khennaoui, A.A.; Zehrour, O.; Bendoukha, S.; Grassi, G.; Pham, V.T. Synchronisation of integer-order and fractional-order discrete-time chaotic systems. *Pramana* **2019**, *92*, 52. [CrossRef]
27. Ouannas, A.; Wang, X.; Khennaoui, A.A.; Bendoukha, S.; Pham, V.T.; Alsaadi, F. Fractional form of a chaotic map without fixed points: Chaos, entropy and control. *Entropy* **2018**, *20*, 720. [CrossRef]
28. Elaydi, S.N. *Discrete Chaos: With Applications in Science and Engineering*; Chapman and Hall/CRC: New York, NY, USA, 2007; ISBN 9780429149191.
29. Khennaoui, A.A.; Ouannas, A.; Bendoukha, S.; Grassi, G.; Wang, X.; Pham, V.T.; Alsaadi, F.E. Chaos, control, and synchronization in some fractional-order difference equations. *Adv. Differ. Equ.* **2019**, *2019*, 1–23. [CrossRef]
30. Ouannas, A.; Khennaoui, A.A.; Bendoukha, S.; Grassi, G. On the Dynamics and Control of a Fractional Form of the Discrete Double Scroll. *Int. J. Bifurc. Chaos.* **2019**, *29*, 1950078. [CrossRef]
31. Ouannas, A.; Khennaoui, A.A.; Odibat, Z.; Pham, V.T.; Grassi, G. On the dynamics, control and synchronization of fractional-order Ikeda map. *Chaos Soliton. Fract.* **2019**, *123*, 108–115. [CrossRef]
32. Ouannas, A.; Khennaoui, A.A.; Grassi, G.; Bendoukha, S. On chaos in the fractional-order Grassi–Miller map and its control. *J. Comput. Appl. Math.* **2019**, 358, 293–305. [CrossRef]
33. Jouini, L.; Ouannas, A.; Khennaoui, A.A.; Wang, X.; Grassi, G.; Pham, V.T. The fractional form of a new three-dimensional generalized Hénon map. *Adv. Differ. Equ.* **2019**, *2019*, 122. [CrossRef]
34. Ouannas, A.; Khennaoui, A.A.; Bendoukha, S.; Vo, T.; Pham, V.T.; Huynh, V. The fractional form of the Tinkerbell map is chaotic. *Appl. Sci.* **2018**, *8*, 2640. [CrossRef]
35. Wu, G.C.; Baleanu, D. Discrete chaos in fractional delayed logistic maps. *Nonlinear Dynam.* **2015**, *80*, 1697–1703. [CrossRef]
36. Jiang, H.; Liu, Y.; Wei, Z.; Zhang, L. A New Class of Two-dimensional Chaotic Maps with Closed Curve Fixed Points. *Int. J. Bifurc. Chaos* **2019**, *29*, 1950094. [CrossRef]

Article

Bifurcations, Hidden Chaos and Control in Fractional Maps

Adel Ouannas [1], Othman Abdullah Almatroud [2], Amina Aicha Khennaoui [3], Mohammad Mossa Alsawalha [2], Dumitru Baleanu [4,5,6], Van Van Huynh [7] and Viet-Thanh Pham [8,*]

1 Laboratory of Mathematics, Informatics and Systems (LAMIS), University of Larbi Tebessi, Tebessa 12002, Algeria; ouannas.adel@univ-tebessa.dz
2 Mathematics Department, Faculty of Science, University of Hail, Hail 81451, Saudi Arabia; o.almatroud@uoh.edu.sa (O.A.A.); m.alswalha@uoh.edu.sa (M.M.A.)
3 Laboratory of Dynamical Systems and Control, University of Larbi Ben M'hidi, Oum El Bouaghi 04000, Algeria; khennaoui.amina@univ-oeb.dz
4 Department of Mathematics, Cankaya University, 06530 Ankara, Turkey; dumitru@cankaya.edu.tr
5 Department of Medical Research, China Medical University Hospital, China Medical University, Taichung 40402, Taiwan
6 Institute of Space Sciences, 76900 Magurele-Bucharest, Romania
7 Modeling Evolutionary Algorithms Simulation and Artificial Intelligence, Faculty of Electrical and Electronics Engineering, Ton Duc Thang University, Ho Chi Minh City 758307, Vietnam; huynhvanvan@tdtu.edu.vn
8 Nonlinear Systems and Applications, Faculty of Electrical and Electronics Engineering, Ton Duc Thang University, Ho Chi Minh City 758307, Vietnam
* Correspondence: phamvietthanh@tdtu.edu.vn

Received: 29 April 2020; Accepted: 25 May 2020; Published: 27 May 2020

Abstract: Recently, hidden attractors with stable equilibria have received considerable attention in chaos theory and nonlinear dynamical systems. Based on discrete fractional calculus, this paper proposes a simple two-dimensional and three-dimensional fractional maps. Both fractional maps are chaotic and have a unique equilibrium point. Results show that the dynamics of the proposed fractional maps are sensitive to both initial conditions and fractional order. There are coexisting attractors which have been displayed in terms of bifurcation diagrams, phase portraits and a 0-1 test. Furthermore, control schemes are introduced to stabilize the chaotic trajectories of the two novel systems.

Keywords: chaos; coexisting attractors; hidden attractors

1. Introduction

Continuous-time and discrete-time chaotic dynamical systems have been extensively studied over the last years. Referring to discrete-time systems number of chaotic maps have been deeply analyzed. Researchers such as Hénon, Lozi and Arnold have attempted to provide different maps with different features. In [1], Hénon proposed the first map by employing the Poincaré section on the Lorenz system. In 1960, the Russian mathematician Vladimer I Arnold discovered a two-dimensional chaotic map using an image of cat [2]. More recently, Lozi developed a new discrete chaotic system by replacing the quadratic term in the Hénon map with quasi-linear term [3]. All of the chaotic attractors in these maps fall into the class of self-excited attractors, for which the initial conditions are located near the unstable fixed points. Recently, another type of attractors called hidden chaotic attractors have been discovered, referring to attractors whose basin of attraction does not intersect with any neighborhoods of any equilibrium of the systems [4]. They are generated by nonlinear systems without equilibrium points or with nonlinear systems with special equilibrium points, for example systems with stable equilibrium points [5]. The topic of chaotic maps with hidden attractors has been recently investigated. For example, in [6] a 1D chaotic discontinuous map without equilibria has been illustrated, whereas in [7] three-dimensional chaotic maps with different types of stable equilibria have been proposed by designing the simplest and most elegant difference systems. Moreover, in [8] 2D chaotic quadratic maps without equilibria and with no discontinuity in the right-hand equations have been introduced. On the basis of the famous Hénon map, some two-dimensional chaotic maps with no equilibrium point and with stable equilibrium have been illustrated in [9]. Some other examples of chaotic attractors with curve equilibrium were numerically presented in [10]. These studies have proven the significant role of hidden attractors in practical engineering applications.

Fractional calculus is a very interesting topic in mathematics with several potential applications in many fields of science and engineering [11]. Recently, several efforts have been devoted to the study of complex dynamics of fractional maps, i.e., maps described by fractional-order difference equations. Researchers have extensively examined the potential application of these maps in many fields such as, engineering, economics and other areas [12–14]. For this purpose, many fractional maps have been reported in the literature to show the different dynamical phenomena. For example, a fractional 3D generalized Hénon map has been studied [15], whereas the Stefanski, Rössler and Wang fractional maps have been illustrated in [16]. Moreover, in [17] the fractional-order version of the Grassi-Miller map is considered, whereas in [18] the dynamics of the fractional discrete double scroll are analyzed in detail. Results show that the dynamics of such systems are clearly dependent on the fractional-order and they are more complex because of their memory effect. Moreover, control and synchronization based on these fractional chaotic maps have also attracted lots of attention [19–22]

Up to the present day, most of the literature on the analysis of fractional maps is still limited to systems with self exited attractors. To our knowledge, fractional maps with hidden attractors have rarely been reported [23], which has inspired researchers to devote themselves to the design of new two and three-dimensional fractional discrete-time chaotic systems [24–26]. Based on the above considerations, new two and three-dimensional fractional chaotic maps with hidden coexisting attractors are developed. The conducted analysis highlights that hidden attractors are generated for some values of the fractional order in the difference equations. The presence of chaos is validated via the bifurcation diagrams and phase portraits. Furthermore, we apply the 0-1 test method to distinguish chaos from regular behavior [27]. The idea under the test is to transform the states of the fractional maps into $p-q$ plots. Generally, unbounded $p-q$ trajectories imply chaotic behavior whereas bounded trajectories imply regular behavior. We also propose two active controllers with the aim of stabilizing the chaotic dynamics of the two fractional maps.

2. Basic Concepts

Herein, some discrete fractional calculus background and theorems are introduced briefly. The definition of fractional Caputo-like difference operator will be given first.

Definition 1. *For a given function $u : N_a = \{a, a+1, ...\} \longrightarrow \mathbb{R}$, the Caputo definition of the fractional difference operator of order $v \notin \mathbb{N}$ is defined as:*

$$^{C}\Delta_a^{v} u(t) = \frac{1}{\Gamma(1-v)} \sum_{s=a}^{t-(1-v)} (t-s-1)^{-v} \Delta_s u(s), \tag{1}$$

where the symbol $\Gamma(.)$ represents the Euler's gamma function and $t \in \mathbb{N}_{a+1-v}$. According to reference [28], the definition of v-th fractional sum of $\Delta_s u(t)$ is expressed in the following:

$$\Delta_a^{-v} u(t) = \frac{1}{\Gamma(v)} \sum_{s=a}^{t-v} (t-s-1)^{(v-1)} u(s), \tag{2}$$

with $t \in \mathbb{N}_{a+v}$ and $v > 0$.

In the following, we need to define the discrete version of the proposed maps. For that, we introduce the following theorem:

Theorem 1. [29] *In particular, for the initial value problem*

$$\begin{cases} {}^{C}\Delta_a^{\nu}u(t) = f(t+\nu-1, u(t+\nu-1)), \\ \Delta^0 u(a) = u_0, \end{cases} \tag{3}$$

the solution turns out to the discrete integral equation as

$$u(t) = u_0(t) + \frac{1}{\Gamma(\nu)} \sum_{s=a+1-\nu}^{t-\nu} (t-s-1)^{(\nu-1)} f(s+\nu-1, u(s+\nu-1)), \tag{4}$$

in which $t \in \mathbb{N}_{\nu+1}$, $\frac{(t-\sigma(s))^{(\nu-1)}}{\Gamma(\nu)}$ *is a discrete kernel function, and* u_0 *is the initial condition.*

Set $a = 0$ *and* $\frac{(t-\sigma(s))^{(\nu-1)}}{\Gamma(\nu)} = \frac{\Gamma(t-s)}{\Gamma(\nu)\Gamma(t-s-\nu+1)}$, *Equation (4) can be transformed to:*

$$u(n) = u_0 + \frac{1}{\Gamma(\nu)} \sum_{j=1}^{n} \frac{\Gamma(n-j+\nu)}{\Gamma(n-j+1)} f(j-1, u(j-1)). \tag{5}$$

Stability of Fractional Order Maps

The stability of equilibrium points for fractional maps can be analysed using the following theorem.

Theorem 2. [30,31] *Let* x_f *be an equilibrium point of a nonlinear fractional difference system* ${}^{C}\Delta_a^{\nu}F(t) = F(x(t+\nu-1))$ *where* $x(t) = (x_1(t), x_2(t), .., x_n(t))^T$, *and* $J(x^f) = \left.\frac{\partial f(x)}{\partial x}\right|_{x=x_f}$ *is the Jacobian matrix at the equilibrium point* x_f. *The equilibrium point* x_f *is asymptotically stable when all the eigenvalues* (λ_i, $i = 1, .., n$) *of* J *verifies:*

$$\lambda_i \in \left\{ z \in \mathbb{C} : \; |z| < \left(2\cos\frac{|\arg z| - \pi}{2-\nu}\right)^{\nu} \text{ and } |\arg z| > \frac{\nu\pi}{2} \right\}. \tag{6}$$

We recall the following lemma, which is a special case of Theorem 2.

Lemma 1. [30] *Two-dimensional fractional map is locally asymptotically stable if* $detJ > 0$ *and*

$$\frac{-Tr(J)}{2} \geq \sqrt{Det(J)}, \quad \text{and} \quad \nu > log_2 \frac{\sqrt{Tr(J)^2 - 4Det(J)} - Tr(J)}{2}, \tag{7}$$

in which $Tr(J)$ *is the trace of the Jacobian matrix* J *and* $Det(J)$ *is the determinant of the matrix* J.

We now present the following theorem, which identifies the asymptotic stability conditions of the zero solution to a linear fractional difference system.

Theorem 3. [30] *For a linear fractional difference system*

$${}^{C}\Delta_a^{\nu}e(t) = \mathbf{M}e(t+\nu-1), \tag{8}$$

where $e(t) = (e_1(t), ..., e_n(t))^T$, $0 < \nu \leq 1$, $\mathbf{M} \in R^{n\times n}$ *and* $\forall t \in \mathbb{N}_{a+1-\nu}$, *the zero equilibrium is asymptotically stable if*

$$\lambda \in \left\{ z \in \mathbb{C}: \ |z| < \left(2\cos \frac{|\arg z| - \pi}{2 - \nu} \right)^{\nu} \ and \ |\arg z| > \frac{\nu \pi}{2} \right\}, \tag{9}$$

for all the eigenvalues λ *of* $\mathbf{M}$.

3. New Two and Three-Dimensional Fractional Maps

3.1. Description of the New Two-Dimensional Fractional Map

An effective method for defining new integer order maps with quadratic nonlinearity terms was proposed by [32]. Motivated by this strategy, we removed and added some terms to obtain the following two-dimensional fractional map:

$$\begin{cases} {}^{C}\Delta_a^{\nu} x(t) = y(t+\nu-1) - x(t+\nu-1), \\ {}^{C}\Delta_a^{\nu} y(t) = -0.33x(t+\nu-1) + Ay(t+\nu-1) - 0.48x^2(t+\nu-1) + 0.47y^2(t+\nu-1) \\ \qquad\qquad +0.01x(t+\nu-1)y(t+\nu-1) - 0.9, \end{cases} \tag{10}$$

with state variables x and y, and system parameter A. $0 < \nu \leq 1$ denotes the fractional order. For calculating the equilibrium points of the fractional map (10), we assign its left hand side to zero:

$$\begin{cases} y = x, \\ -0.33x + Ay - 0.48x^2 + 0.47y^2 + 0.01xy - 0.9 = 0, \end{cases} \tag{11}$$

from system of Equation (11) it follows:

$$(-0.33 + A)x - 0.9 = 0. \tag{12}$$

It is easy to verify that the fractional map (10) has a unique equilibrium point when $A \neq 0.33$. The Jacobian matrix of the fractional map (10) at an arbitrary point (x, y), is given by:

$$J = \begin{pmatrix} -1 & 1 \\ -0.33 - 1.29x + 0.01y & A + 0.94y + 0.01x \end{pmatrix}. \tag{13}$$

The associated characteristic equation is defined in terms of the trace ($Tr(J)$) and determinant ($Det(J)$) of the matrix J by:

$$det(\lambda I - J) = \lambda^2 - Tr(J)\lambda + Det(J) = 0, \tag{14}$$

where $Tr(J) = -1 + A + 0.94y + 0.01x$ and $Det(J) = -A + 0.3 - 0.95y + 1.28x$. For $A = -0.83$, the fractional map has a unique equilibrium point $E = (\frac{-0.9}{1.16}, \frac{-0.9}{1.16})$. Based on Lemma 1 the equilibrium point E is stable when $\nu > log_2 \frac{\sqrt{Tr(J)^2 - 4Det(J)} - Tr(J)}{2}$. By simple calculation it is easy to verify that the equilibruim point E is asymptotically stable when the fractional order $\nu > 0.1430$.

In order to investigate the variety of dynamics behavior that can be observed in the fractional map (10) near to the stable equilibrium point E, it is important to present at first the corresponding numerical formula. In view of Theorem 1, the fractional map (10) is changed to:

$$\begin{cases} x(n) = x_0 + \frac{1}{\Gamma(\nu)} \sum\limits_{j=1}^{n} \frac{\Gamma(n-j+\nu)}{\Gamma(n-j+1)} (y(j-1) - x(j-1)), \\ y(n) = y_0 + \frac{1}{\Gamma(\nu)} \sum\limits_{j=1}^{n} \frac{\Gamma(n-j+\nu)}{\Gamma(n-j+1)} (-0.33x(j-1) + Ay(j-1) - 0.48x^2(j-1) + \\ \qquad\qquad 0.47y^2(j-1) + 0.01x(j-1)y(j-1) - 0.9), \end{cases} \tag{15}$$

where x_0 and y_0 are the initial conditions. Numerical analysis are presented in the next subsection.

3.2. Bifurcation and 0-1 Test

In the following, the coexisting of hidden attractors in the fractional map are revealed by phase portraits, bifurcation diagrams and 0–1 test. The phase portrait is a geometric representation of the trajectories of a dynamical system. For the system parameter $A = -0.83$, fractional order $\nu = 0.999$ and initial condition $(x_0, y_0) = (0.32, -1.85)$, a hidden strange attractor is numerically obtained as it observed in Figure 1. The Lyapunov exponents (LEs) of the fractional map (10) are $LE_1 = 0.0107$. $LE_2 = -0.0279$. Since the maximum Lyapunov exponent is larger than zero, we can determine that the hidden attractor is chaotic.

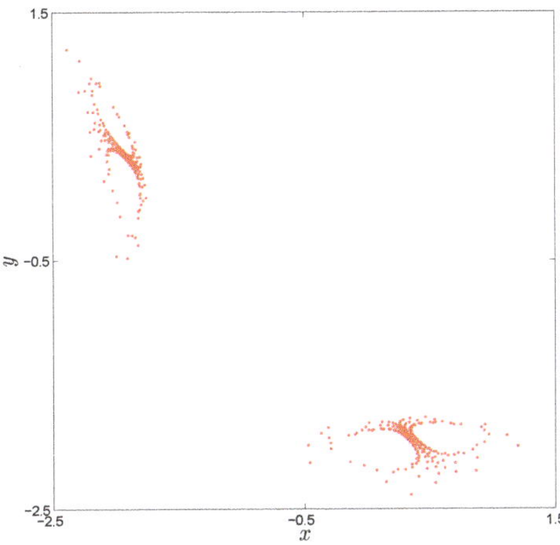

Figure 1. Strange attractor of the two-dimensional fractional map (10) for $\nu = 0.999$ and $A = -0.83$.

For $A = -0.83$, the dynamic evolution of the fractional map versus ν is given via plotting its bifurcation diagram (see Figure 2). The bifurcation diagram is obtained by plotting the local maximum value of the variable x for two sets of initial conditions. The blue diagram begins with the initial condition $(x_0, y_0) = (0.32, -1.85)$ and the red diagram begins with the initial condition $(x_0, y_0) = (-0.32, -1.85)$. When the fractional order ν varies from 1 to 0.999, our fractional system (10) generates chaos with transient states. As ν decreases further a coexisting periodic orbits are obtained. The coexisting attractors with different values of ν are shown in Figure 3. Two periodic attractors coexist for $\nu = 0.9989$, $\nu = 0.9987$, $\nu = 0.9984$ with initial values $(x_0, y_0) = (0.32, -1.85)$ and $(x_0, y_0) = (-0.32, -1.85)$ as shown in Figure 3b–d. A hidden chaotic attractor is observed with order $\nu = 0.9996$. It is noticed that the type of hidden attractors not only depend on the value of ν but also on the initial conditions.

Now, the $p - q$ plots of the 0-1 test are used to confirm the property of coexisting attractors. The 0-1 test is relatively new method that was proposed by Gottwald and Melbourne [27] to test the presence of chaos in a series of data which originate from deterministic systems. For the fractional map (10) consider a set of discrete points $x(j)$ where $j = 1, ..., N$. For a randomly chosen constant $c \in (0, \pi)$, decompose the state $x(n)$ into two components p and q as:

$$p(n) = \sum_{j=1}^{n} x(j)\cos(jc), \quad q(n) = \sum_{j=1}^{n} x(j)\sin(jc), \quad n = 1, 2, .., N. \tag{16}$$

Generally, unbounded $p - q$ trajectories imply chaotic behavior whereas bounded trajectories imply regular behavior. As in Figure 3 we chose to fix the system parameter A to $A = -0.83$ and vary the fractional order ν. Figure 4 shows the $p - q$ plots with $\nu = 0.9996$ and $\nu = 0.9984$, where the blue plots are obtained for the initial values $(x_0, y_0) = (0.32, -1.85)$ and the red plots are obtained for $(x_0, y_0) = (-0.32, -1.85)$. In particular, Figure 4a shows Brownian like trajectories in p versus q plan for both initial conditions, confirming that the dynamics of the fractional map (10) are chaotic for both initial values and fractional order $\nu = 0.9996$. When $\nu = 0.9984$, Figure 4b depicts bounded like trajectories in p versus q plane for both initial conditions, which confirms the coexisting of hidden periodic orbits.

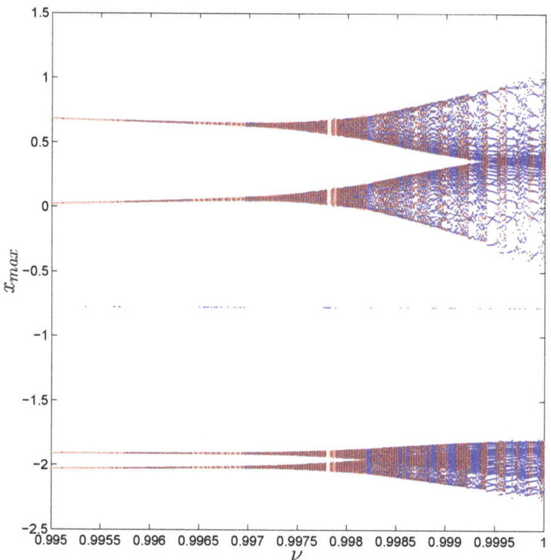

Figure 2. Bifurcation diagrams of the two-dimensional fractional map (10) versus ν for $A = -0.83$.

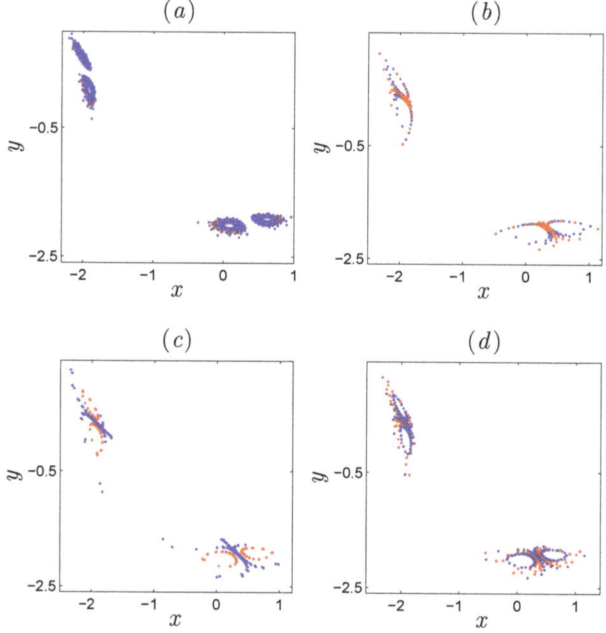

Figure 3. The coexisting hidden attractors of the two-dimensional fractional map (10) for system parameter $A = -0.83$ and initial condition $(-0.32, -1.85)$ for red attractors and $(0.32, -1.85)$ for blue attractors, with fractional order varying: (**a**) $\nu = 0.9996$, (**b**) $\nu = 0.9989$, (**c**) $\nu = 0.9987$, (**d**) $\nu = 0.9984$.

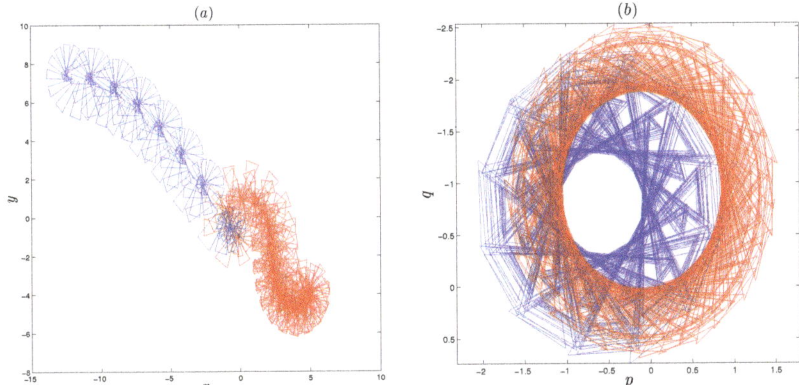

Figure 4. The 0-1 test of the two-dimensional fractional map. (**a**) Brownian like trajectories for both initial conditions with $\nu = 0.9996$, (**b**) bounded trajectories for both initial conditions with $\nu = 0.9984$.

3.3. Description of the New Three-Dimensional Fractional Map

We introduce here a new three-dimensional fractional map with two nonlinearities:

$$\begin{cases} {}^{C}\Delta_a^{\nu}x(t) = y(t+\nu-1) - x(t+\nu-1), \\ {}^{C}\Delta_a^{\nu}y(t) = z(t+\nu-1) - y(t+\nu-1), \\ {}^{C}\Delta_a^{\nu}z(t) = -y(t+\nu-1) - 0.4z(t+\nu-1) - 0.1x(t+\nu-1)z(t+\nu-1) \\ \qquad\qquad +0.1y(t+\nu-1)z(t+\nu-1) + 1, \end{cases} \tag{17}$$

where ν is the fractional order in which $0 < \nu \leq 1$. The equilibrium points of the fractional map (17) are found by:

$$\begin{cases} y = x, \\ z = y, \\ -y - 0.4z - 0.1xz + 0.1yz + 1 = 0, \end{cases} \tag{18}$$

from Equation (18) we have the simplified form

$$-1.4x + 1 = 0. \tag{19}$$

It is obvious that the point $E_f = (\frac{1}{1.4}, \frac{1}{1.4}, \frac{1}{1.4})$ is the only equilibrium point and it is stable. Similarly, the discrete version of system (17) is obtained by applying Theorem 1 as follows:

$$\begin{cases} x(n) = x_0 + \frac{1}{\Gamma(\nu)} \sum\limits_{j=1}^{n} \frac{\Gamma(n-j+\nu)}{\Gamma(n-j+1)} (y(j-1) - x(j-1)), \\ y(n) = y_0 + \frac{1}{\Gamma(\nu)} \sum\limits_{j=1}^{n} \frac{\Gamma(n-j+\nu)}{\Gamma(n-j+1)} (z(j-1) - y(j-1)), \\ z(n) = z_0 + \frac{1}{\Gamma(\nu)} \sum\limits_{j=1}^{n} \frac{\Gamma(n-j+\nu)}{\Gamma(n-j+1)} (-y(j-1) - 0.4z(j-1) \\ \qquad -0.1x(j-1)z(j-1) + 0.1y(j-1)z(j-1) + 1). \end{cases} \tag{20}$$

Here, x_0, y_0 and z_0 are the initial states. Taking advantage of the numerical solution (20), numerical simulation can be performed to show the basic properties of the novel system (17).

3.4. Bifurcation and 0-1 Test

To evaluate the dynamic properties of the new system, the initial condition need to set as $(x_0, y_0, z_0) = (-0.26, 3.83, -2.22)$. The three-dimensional fractional map (17) with stable equilibrium point display strange attractor for $\nu = 0.999$ as shown in Figure 5. Here, the largest Lyapunove exponent (LE_{max}) of the fractional map (17) is calculated as $LE_{max} = 0.0488$, so the strange attractor in Figure 5 is chaotic.

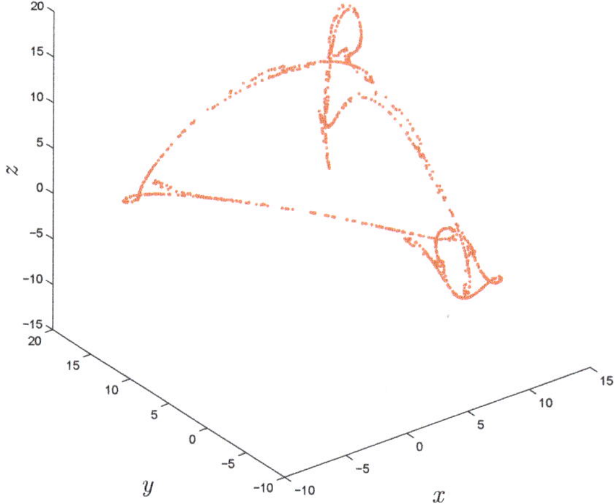

Figure 5. Strange attractor of the three-dimensional fractional map (17).

The bifurcation diagram can help us to observe the dynamic behaviours of the fractional map (17). The fractional order ν is considered as the only bifurcation parameter in this subsection. Changing ν from 0.95 to 1 and setting two different initial conditions we can obtain the bifurcation diagram in Figure 6, where the blue color diagram and the red color diagram are yielded from the initial conditions $(x_0, y_0, z_0) = (-0.26, -3.83, -2.22)$ and $(x_0, y_0, z_0) = (-0.26, 3.83, -2.22)$, respectively. When $\nu \in (0.9503, 0.9922)$ coexisting fixed point attractors are observed. When the fractional order increases from 0.9922 to 0.9975, we can observe coexisting periodic windows. For $0.9957 < \nu \leq 1$, the states of the fractional map go from periodic behavior to chaos. The coexistence of the different dynamic behaviours of the new fractional map (17) is confirmed with phase portrait and $p - q$ plots of the 0-1 test for different values of ν. Figure 7 shows the phase portraits of the system while Figure 8 shows the $p - q$ plots for the same fractional orders where the blue plots are obtained for the initial values $(x_0, y_0, z_0) = (-0.26, -3.83, -2.22)$ and the red plots are obtained for $(x_0, y_0, z_0) = (-0.26, 3.83, -2.22)$. Figure 8a depicts bounded like trajectories, which confirms the periodic phenomena of the hidden attractor in Figure 7a for $\nu = 0.993$. For $\nu = 0.9984$, a hidden attractor coexists as shown in Figure 7, whose Brownian-like trajectories can be seen in Figure 8.

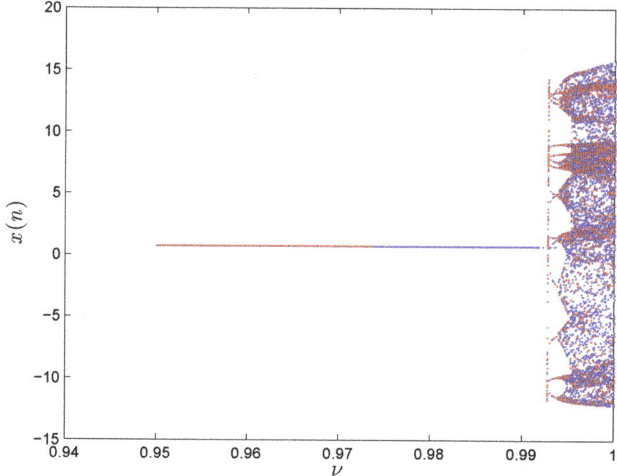

Figure 6. Bifurcation diagram of the three-dimensional fractional map (15) versus ν.

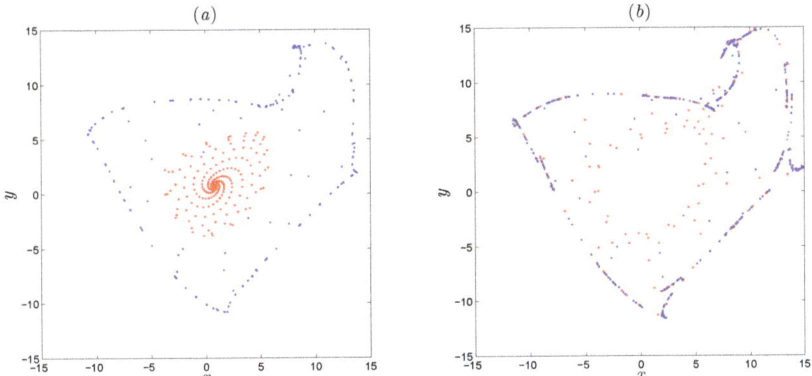

Figure 7. Coexisting hidden attractors of the three-dimensional fractional map (17) with initial condition $(-0.26, 3.83, -2.22)$ for red attractors and $(-0.26, -3.83, -2.22)$ for blue attractors, (**a**) for fractional order $\nu = 0.993$, (**b**) for fractional order $\nu = 0.9984$.

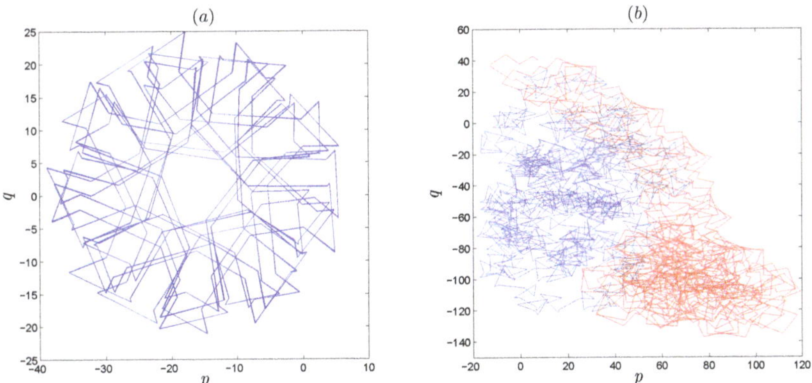

Figure 8. The $p - q$ plots of the three-dimensional fractional map, (**a**) bounded trajectories for $\nu = 0.993$, (**b**) Brownian-like trajectories for both initial conditions with $\nu = 0.9984$.

4. Chaos Control

In this section, adaptive controllers are designed to stabilize the chaotic trajectories of two-dimensional fractional map (10) as well as the three-dimensional fractional map (17) to zero asymptotically. The purpose of such controllers is to assure that all states of the proposed maps converge to zero, by adding an adaptive term into one of the systems states that forces the systems to become linear. A purely linear fractional difference system of the form ${}^{C}\Delta_{a}^{\nu}F(t) = MF(t+\nu-1)$ has zero as its equilibrium. We use the stability theory of linear fractional difference systems to guarantee the asymptotic stability of the zero equilibrium.

The controlled two-dimensional fractional map (10) is given by

$$\begin{cases} {}^{C}\Delta_{a}^{\nu}x(t) = y(t+\nu-1) - x(t+\nu-1), \\ {}^{C}\Delta_{a}^{\nu}y(t) = -0.33x(t+\nu-1) + Ay(t+\nu-1) - 0.48x^{2}(t+\nu-1) + 0.47y^{2}(t+\nu-1) \\ \qquad\qquad +0.01x(t+\nu-1)y(t+\nu-1) - 0.9 + \mathbf{C}(t), \end{cases} \tag{21}$$

where $\mathbf{C}(t)$ denotes the one-dimensional controller. Our goal is to find a suitable one-dimensional controller such that both of states of system are stabilized towards zero asymptotically. For that, we propose the following theorem.

Theorem 4. *The two-dimensional fractional map (10) can be stabilized by the one-dimensional controller described by*

$$\mathbf{C}(t) = 0.33x(t) + 0.48x^2(t) + 0.47y^2(t) - 0.01x(t)y(t) + 0.9, \qquad t \in \mathbb{N}_{a+\nu}. \tag{22}$$

Proof. If we substitute the adaptive control law (26) to the second state of the fractional map we obtain the following set of equations:

$$\begin{cases} {}^{C}\Delta_a^{\nu}x(t) = -x(t+\nu-1) + y(t+\nu-1), \\ {}^{C}\Delta_a^{\nu}y(t) = Ay(t+\nu-1). \end{cases} \tag{23}$$

System (23) can be represent by the following compact form:

$${}^{C}\Delta_a^{\nu}(x,y)^T = \mathbf{M}\times(x(t-1+\nu), y(t-1+\nu))^T, \tag{24}$$

where

$$\mathbf{M} = \begin{pmatrix} -1 & 1 \\ 0 & A \end{pmatrix}. \tag{25}$$

So, the eigenvalues λ_1, λ_2 of the matrix $\mathbf{M}$ have been found that satisfies the condition of the Theorem 3:

$$|\arg \lambda_i| = \pi > \frac{\nu\pi}{2} \text{ and } |\lambda_i| \leq 1 \leq \left(2\cos\frac{|\arg \lambda_i| - \pi}{2-\nu}\right)^{\nu}, \quad i = 1,2.$$

Thus the zero equilibrium of (24) is asymptotically stable, therefore, we can conclude that the proposed two-dimensional system (10) is stabilized. □

Now, we give the evolution of states and phase space plots of the controlled system to confirm the above theoretical results. In Figure 9, the value parameter is taken as $A = -0.83$ and the fractional order is chosen as $\nu = 0.999$. The simulation are done using the initial value $(x_0, y_0) = (0.32, -1.85)$. It is clear that the controller has compelled the states towards zeros.

With the same procedure, we may state the following result regarding the chaos control of the three-dimensional fractional map (17).

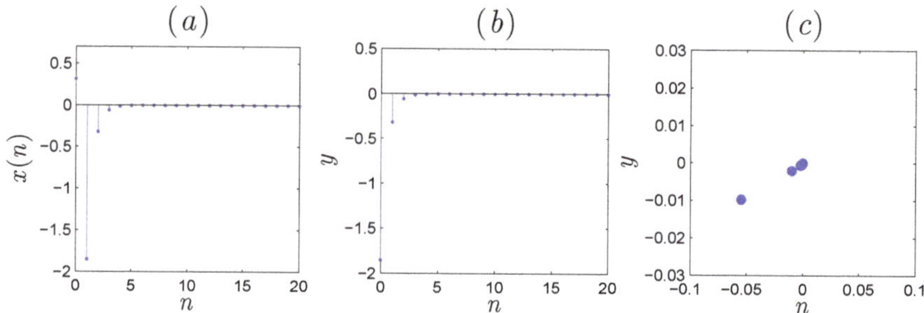

Figure 9. (**a**) Stabilization of state $x(n)$, (**b**) stabilization of state $y(n)$, (**c**) attractor of the controlled system (10) for $\nu = 0.999$.

Theorem 5. *The three-dimensional fractional map (17) can be stabilized by the one-dimensional controller described by*

$$\mathbf{L}(t) = y(t) + 0.1z(t)(x(t) - y(t)) - 1, \qquad t \in \mathbb{N}_{a+\nu}. \tag{26}$$

Again, assuming the fractional order value $\nu = 0.998$, the resulting states and phase plot are depicted in Figure 10. The results confirms the success of the proposed law in stabilizing the systems states asymptotically.

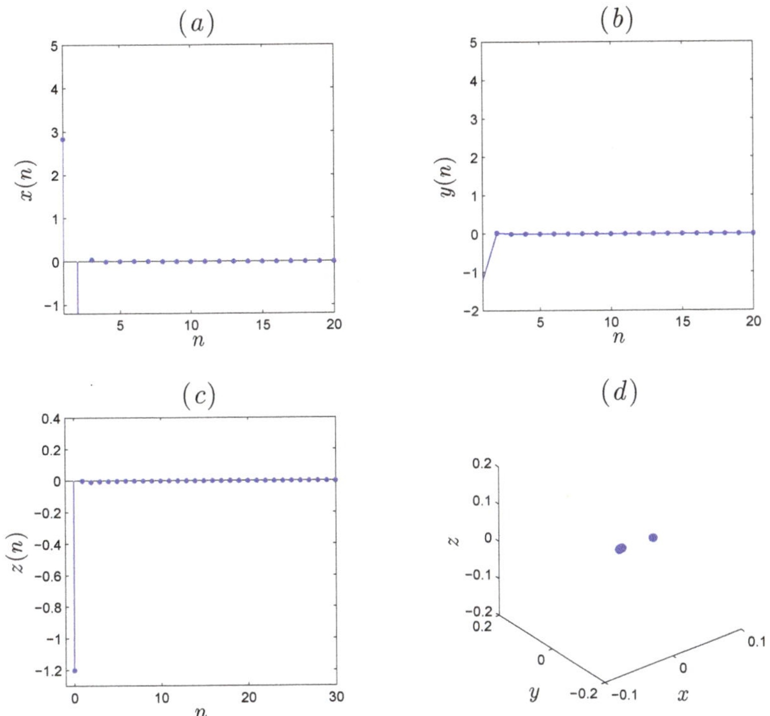

Figure 10. (**a**) Stabilization of state $x(n)$, (**b**) stabilization of state $y(n)$, (**c**) stabilization of state $z(n)$,(**d**) attractor of the controlled system (17) for $\nu = 0.998$.

5. Conclusions

In this paper, two fractional maps with stable equilibrium points are developed. The complex dynamics of these fractional maps are discussed numerically with some changes in system parameters and fractional order ν. The presence of chaos and the property of coexisting attractors have been validated via the computation of a 0-1 test and bifurcation diagrams. The type of hidden attractors does not only depend on the value of fractional order but also on the initial condition. Finally, one-dimensional control laws have been designed, with the aim to stabilize at zero the dynamics of the two proposed maps. These novel maps are good candidates for engineering applications. We will focus on putting this claim to the test in a future study. An interesting question under investigation is that of developing a new type of fractional pseudo number generator based on the fractional maps with hidden attractors for potential cryptographic applications. It is our intention to investigate this issue further in future studies.

Author Contributions: Conceptualization, A.O. and D.B.; Data curation, O.A.A. and A.A.K.; formal analysis, O.A.A. and A.A.K.; funding acquisition, V.V.H.; investigation, A.O.; methodology, M.M.A. and D.B.; project administration, A.O.; resources, M.M.A. and V.-T.P.; software, V.-T.P.; supervision, D.B.; visualization, O.A.A., M.M.A. and V.V.H.; writing—original draft, A.A.K.; writing—review and editing, V.V.H. and V.-T.P. All authors have read and agreed to the published version of the manuscript.

Funding: This research received no external funding.

Acknowledgments: This work has been supported by Scientific Research Deanship at University of Ha'il, Saudi Arabia, though Project Number RG-191307. The author Adel Ouannas was supported by the Directorate General for Scientific Research and Technological Development of Algeria.

Conflicts of Interest: The authors declare no conflict of interest.

References

1. Hénon, M. A two-dimensional mapping with a strange attractor. In *The Theory of Chaotic Attractors*; Springer: New York, NY, USA, 2004; pp. 94–102.
2. Anton, H.; Rorres C. *Elementary Linear Algebra: Application Version*, 7th ed.; Howard, Drexel Unversity: Philadelphia, PA, USA, 1994; pp. 571–572.
3. Lozi R . Un attracteur étrange du type attracteur de Hénon. *J. Phys. Colloq.* **1978**, *39*, C5-9.
4. Leonov, G.A.; Kuznetsov, N.V. Hidden attractors in dynamical systems: from hidden oscillation in Hilbert–Kolmogorov, Aizerman and Kalman problems to hidden chaotic attractor in Chua circuits. *Int. J. Bifurc. Chaos* **2013**, *23*, 1330002. [CrossRef]
5. Jafari, S.; Sprott, J.C.; Nazarimehr, F. Recent new examples of hidden attractors *Eur. Phys. J. Spec. Top.* **2015**, *224*, 1469. [CrossRef]
6. Jafari, S.; Pham, V.T.; Golpayegani, S.M.R.H.; Moghtadaei, M.; Kingni, S.T. The relationship between chaotic maps and some chaotic systems with hidden attractors. *Int. J. Bifurc. Chaos* **2016**, *26*, 1650211. [CrossRef]
7. Jiang, H.; Liu, Y.; Wei, Z.; Zhang, L. A new class of three-dimensional maps with hidden chaotic dynamics. *Int. J. Bifurc. Chaos* **2016**, *26*, 1650206. [CrossRef]
8. Shabestari, P.S.; Panahi, S.; Hatef, B.; Jafari, S.; Sprott, J.C. Two simplest quadratic chaotic maps without equilibrium. *Int. J. Bifurc. Chaos* **2018**, *28*, 1850144.
9. Jiang, H.; Liu, Y.; Wei, Z.; Zhang, L. Hidden chaotic attractors in a class of two-dimensional maps. *Nonlinear Dyn.* **2016**, *85*, 2719–2727.
10. Jiang, H.; Liu, Y.; Wei, Z.; Zhang, L. A New Class of Two-Dimensional Chaotic Maps with Closed Curve Fixed Points. *Int. J. Bifurc. Chaos* **2019**, *29*, 1950094. [CrossRef]
11. Wu, G.C.; Baleanu, D. Discrete fractional logistic map and its chaos. *Nonlinear Dyn.* **2014**, *75*, 283–287. [CrossRef]
12. Khennaoui, A.A.; Ouannas, A.; Bendoukha, S.; Wang, X.; Pham, V.T. On chaos in the fractional-order discrete-time unified system and its control synchronization. *Entropy* **2018**, *20*, 530. [CrossRef]
13. Kang, X.; Luo, X.; Zhang, X.; Jiang, J. Homogenized Chebyshev-Arnold map and its application to color image encryption. *IEEE Access* **2019**, *7*, 114459–114471. [CrossRef]
14. Xin, B.; Peng, W.; Kwon, Y. A fractional-order difference Cournot duopoly game with long memory. *arXiv* **2019**, arXiv: 1903.0430.
15. Jouini, L.; Ouannas, A.; Khennaoui, A.A.; Wang, X.; Grassi, G.; Pham, V.T. The fractional form of a new three-dimensional generalized Hénon map. *Adv. Differ. Equ.* **2019**, *1*, 122. [CrossRef]
16. Khennaoui, A.A.; Ouannas, A.; Bendoukha, S.; Grassi, G.; Lozi, R.P.; Pham, V.T. On fractional–order discrete–time systems: Chaos, stabilization and synchronization. *Chaos Solitons Fractals* **2019**, *119*, 150–162. [CrossRef]
17. Ouannas, A.; Khennaoui, A.A.; Grassi, G.; Bendoukha, S. On chaos in the fractional-order Grassi–Miller map and its control. *J. Comput. Appl. Math.* **2019**, *358*, 293–305. [CrossRef]
18. Ouannas, A.; Khennaoui, A.A.; Bendoukha, S.; Grassi, G. On the dynamics and control of a fractional form of the discrete double scroll. *Int. J. Bifurc. Chaos* **2019**, *29*, 1950078. [CrossRef]
19. Ouannas, A.; Khennaoui, A.A.; Odibat, Z.; Pham, V.T.; Grassi, G. On the dynamics, control and synchronization of fractional-order Ikeda map. *Chaos Solitons Fractals* **2019**, *123*, 108–115. [CrossRef]
20. Ouannas, A.; Khennaoui, A.A.; Grassi, G.; Bendoukha, S. On the Q–S chaos synchronization of fractional-order discrete-time systems: General method and examples. *Discret. Dyn. Nat. Soc.* **2018**, *2018*, 2950357. [CrossRef]
21. Khennaoui, A.A.; Ouannas, A.; Bendoukha, S.; Grassi, G.; Wang, X.; Pham, V.T. Generalized and inverse generalized synchronization of fractional-order discrete-time chaotic systems with non-identical dimensions. *Adv. Differ. Equ.* **2018**, *2018*, 1–14.

22. Bendoukha, S.; Ouannas, A.; Wang, X.; Khennaoui, A.A.; Pham, V.T.; Grassi, G.; Huynh, V.V. The Co-existence of different synchronization types in fractional-order discrete-time chaotic systems with non–identical dimensions and orders. *Entropy* **2018**, *20*, 710. [CrossRef]
23. Ouannas, A.; Wang, X.; Khennaoui, A.A.; Bendoukha, S.; Pham, V.T.; Alsaadi, F.E. Fractional form of a chaotic map without fixed points: Chaos, entropy and control. *Entropy* **2018**, *20*, 720.
24. Khennaoui, A.A.; Ouannas, A.; Boulaaras, S.; Pham, V.T.; Taher Azar, A. A fractional map with hidden attractors: Chaos and control. *Eur. Phys. J. Spec. Top.* **2020**, *229*, 1083–1093.
25. Ouannas, A.; Khennaoui, A.A.; Momani, S.; Grassi, G.; Pham, V.T. Chaos and control of a three-dimensional fractional order discrete-time system with no equilibrium and its synchronization. *AIP Adv.* **2020**, *10*, 045310.
26. Ouannas, A.; Khennaoui, A.; Momani, S.; Pham, V.T.; El-Khazali, R. Hidden attractors in a new fractional-order discrete system: Chaos, complexity, entropy and control. *Chin. Phys. B* **2020**, *29*, 050504.
27. Gottwald, G.A.; Melbourne, I. On the implementation of the 0–1 test for chaos. *SIAM J. Appl. Dyn. Syst.* **2009**, *8*, 129–145. [CrossRef]
28. Atici, F.M.; Eloe, P.W. Discrete fractional calculus with the nabla operator. *Electron. J. Qual. Theory Differ. Equ.* **2009**, *3*, 1–12. [CrossRef]
29. Anastassiou, G.A. Principles of delta fractional calculus on time scales and inequalities. *Math. Comput. Model.* **2010**, *52*, 556–566. [CrossRef]
30. Cermak, J.; Gyori, I.; Nechvatal, L. On explicit stability conditions for a linear fractional difference system. *Fract. Calc. Appl. Anal.* **2015**, *18*, 651–672. [CrossRef]
31. Mozyrska, D.; Wyrwas, M. Stability by linear approximation and the relation between the stability of difference and differential fractional systems. *Math. Methods Appl. Sci.* **2017**, *40*, 4080–4091. [CrossRef]
32. Sprott, J.C. *Strange Attractors: Creating Patterns in Chaos*; M & T Books: New York, NY, USA, 1993.

Article

Generalized Bessel Polynomial for Multi-Order Fractional Differential Equations

Mohammad Izadi [1,*] and Carlo Cattani [2]

1 Department of Applied Mathematics, Faculty of Mathematics and Computer, Shahid Bahonar University of Kerman, Kerman 76169-14111, Iran
2 Engineering School, DEIM, University of Tuscia, 01100 Viterbo, Italy; cattani@unitus.it
* Correspondence: izadi@uk.ac.ir

Received: 19 May 2020; Accepted: 6 July 2020; Published: 30 July 2020

Abstract: The main goal of this paper is to define a simple but effective method for approximating solutions of multi-order fractional differential equations relying on Caputo fractional derivative and under supplementary conditions. Our basis functions are based on some original generalization of the Bessel polynomials, which satisfy many properties shared by the classical orthogonal polynomials as given by Hermit, Laguerre, and Jacobi. The main advantages of our polynomials are two-fold: All the coefficients are positive and any collocation matrix of Bessel polynomials at positive points is strictly totally positive. By expanding the unknowns in a (truncated) series of basis functions at the collocation points, the solution of governing differential equation can be easily converted into the solution of a system of algebraic equations, thus reducing the computational complexities considerably. Several practical test problems also with some symmetries are given to show the validity and utility of the proposed technique. Comparisons with available exact solutions as well as with several alternative algorithms are also carried out. The main feature of our approach is the good performance both in terms of accuracy and simplicity for obtaining an approximation to the solution of differential equations of fractional order.

Keywords: caputo fractional derivative; bessel functions; collocation method; multi-order fractional differential equations

JEL Classification: 26A33; 65L60; 42C05; 65L05

---

1. Introduction

In recent years, fractional calculus has becoming an efficient and successful tool for the analysis of several physical-mathematical problems. The main reason for the increasing number of papers dealing with fractional problems is also explained by the intrinsic and natural possibility of the fractional calculus to take into account also some memory effects, which is quite impossible by using the ordinary differential operators [1]. In this work, we consider the nonlinear multi-order fractional differential equations (MOFDEs) of the form

$$\mathcal{D}_{\star}^{\gamma}x(t) = F\left(t, x(t), \mathcal{D}_{\star}^{\beta_1}x(t), \ldots, \mathcal{D}_{\star}^{\beta_\ell}x(t)\right), \quad 0 \leqslant t \leqslant L, \tag{1}$$

subjected by the following boundary or supplementary conditions

$$H_j(x(\eta_j), x^{(1)}(\eta_j), \ldots, x^{(m-1)}(\eta_j)) = d_j \quad j = 0, 1, \ldots, m-1, \tag{2}$$

where H_j are linear functions, $d_j \in \mathbb{R}$, and η_j are some given points in $[0, L]$. In (1), $\mathcal{D}_{\star}^{\gamma}$ denotes the Caputo fractional derivative operator such that $m-1 < \gamma \leqslant m$, $m \in \mathbb{N}$, and $0 < \beta_1 < \beta_2 < \ldots < \beta_\ell < \gamma$

are real constants. The function F can be either linear or nonlinear function of its arguments. In [2], some preliminary results both on the existence and uniqueness of the solution of MOFDEs (1) are obtained.

It is well-known that usually the exact solution of fractional differential equations cannot be obtained analytically. Therefore, many authors have recently developed some suitable numerical methods for such equations. Among the many approximation algorithms for (1) and (2), we mention the systems-based decomposition approach [3], the Adomian decomposition method [4], the spectral methods [5–8], the B-spline approach [9], and the generalized triangular function [10].

It is known that the traditional orthogonal polynomials such as Jacobi, Hermit, and Laguerre are solution of a second order differential equation. In addition, the derivatives of these polynomials constitute an orthogonal system. Moreover, there exist another set of polynomials with the two aforementioned properties. They satisfy the following second order differential equation

$$x^2 y''(x) + 2(x+1) y'(x) - n(n+1) y(x) = 0, \tag{3}$$

where n is a positive integer. Krall and Frink [11] called these the Bessel polynomials thank to their close relation with the Bessel functions of half-integral order. In fact, they have shown that these polynomials occur naturally as the solutions of the classical wave equation in spherical coordinates. These polynomials also appear in the study of various mathematical topics including transcendental number theory [12,13] and student t-distributions [14]. These polynomials seem to have been considered first by Bochner [15] as pointed in Grosswald [16]. However, Krall and Frink considered them in more general setting so that to include also the polynomial solutions of the differential Equation (3). The properties of Bessel polynomials such as recurrence relations, generating functions, and orthogonality were investigated in [11]. The algebraic properties of these polynomials were considered by Grosswald [16]. Some more information about the theory and applications of Bessel polynomials can be found e.g., in [17].

In this research, we establish a new approximation algorithm based upon the Bessel polynomials to obtain a solution of a fractional model (1). In fact, one of our motivation comes from a recent paper [18], which proved the total positiveness of any collocation matrix of theses polynomials evaluated at positive (collocation) points. To the best of our knowledge, this is the first attempt to study these polynomials for approximating MOFDEs. In summary, the main idea behind the presented approximation algorithm based on using the Bessel polynomials with together the collocation points is that it transforms the differential operators in (1) to an equivalent algebraic form, thus greatly reducing the numerical efforts. It should be mentioned that our Bessel polynomials are different from those Bessel functions known as Bessel functions of the first kind that previously considered in the literature, see [19] for a recent review.

The content of the paper is structured as follows. In Section 2 some relevant properties of the Caputo fractional derivative and the generalized Taylor's formula for the Caputo derivative are presented. Section 3 is dedicated to the definitions of Bessel polynomials and their generalized fractional-order counterpart. Moreover, the results about the convergence and error bound of these polynomials are established. In Section 4, where a collocation method also shown to solve the MOFDEs. By using these Bessel basis functions along with collocation points, the proposed method converts the MOFDEs into a nonlinear matrix equation. Hence, the residual error function is introduced to assess the accuracy of Bessel-collocation scheme when the exact solutions are not available. In Section 5, some examples with various parameters together with error evaluation are given to show the utility and applicability of the method. The obtained results are interpreted through tables and figures. Finally, in Section 6, the report ends with a summary and conclusion.

2. Some Preliminaries

To continue, some definitions and theorems from fractional calculus theory are presented.

Definition 1. *Let $f(t)$ be a m-times continuously differentiable function. The fractional derivative $\mathcal{D}_\star^q$ of $f(t)$ of order $q > 0$ in the Caputo's sense is defined as*

$$\mathcal{D}_\star^q f(t) = \begin{cases} \mathcal{I}^{m-q} f^{(m)}(t) & \text{if } m-1 < q < m, \\ f^{(m)}(t), & \text{if } q = m, \quad m \in \mathbb{N}, \end{cases} \tag{4}$$

where

$$\mathcal{I}^q f(t) = \frac{1}{\Gamma(q)} \int_0^t \frac{f(s)}{(t-s)^{1-q}} ds, \quad t > 0.$$

The properties of the operator $\mathcal{D}_\star^q$ can be found in [1]. Besides the linearity, the following properties will be also used

$$\mathcal{D}_\star^q(C) = 0 \quad (C \text{ is a constant}), \tag{5}$$

$$\mathcal{D}_\star^q t^\beta = \begin{cases} \frac{\Gamma(\beta+1)}{\Gamma(\beta+1-q)} t^{\beta-q}, & \text{for } \beta \in \mathbb{N}_0 \text{ and } \beta \geqslant \lceil q \rceil, \text{ or } \beta \notin \mathbb{N}_0 \text{ and } \beta > \lfloor q \rfloor, \\ 0, & \text{for } \beta \in \mathbb{N}_0 \text{ and } \beta < \lceil q \rceil. \end{cases} \tag{6}$$

Now, we define a generalization of Taylor's formula which involves Caputo fractional derivatives (see a proof in [20]).

Theorem 1 (Generalized Taylor's formula). *Assuming that $\mathcal{D}_\star^{k\alpha} g(x) \in C(0, L]$, where $k = 0, 1, \ldots, N$, $0 < \alpha \leqslant 1$, and $L > 0$. Then, there exists a $0 < \theta \leqslant x$ such that*

$$g(x) = \sum_{j=0}^{N-1} \frac{x^{j\alpha}}{\Gamma(j\alpha+1)} \mathcal{D}_\star^{j\alpha} g(0^+) + \frac{x^{N\alpha}}{\Gamma(N\alpha+1)} \mathcal{D}_\star^{N\alpha} g(\theta), \quad \forall x \in [0, L].$$

Also, we have

$$\left| g(x) - \sum_{j=0}^{N-1} \frac{x^{j\alpha}}{\Gamma(j\alpha+1)} \mathcal{D}_\star^{j\alpha} g(0^+) \right| \leqslant \frac{x^{N\alpha}}{\Gamma(N\alpha+1)} M^\alpha,$$

where $|\mathcal{D}_\star^{N\alpha} g(\theta)| \leqslant M^\alpha$ and $\mathcal{D}_\star^{N\alpha} = \mathcal{D}_\star^\alpha \cdot \mathcal{D}_\star^\alpha \cdots \mathcal{D}_\star^\alpha$ (N-times).

Finally, we define the concept of the weighted norm used in the proof of Theorem 2:

Definition 2. *Let assume that $g \in C(0, L]$ and $w(t)$ is a weight function. Then*

$$\|g(t)\|_w = \left(\int_0^L |g(t)|^2 w(t) dt \right)^{\frac{1}{2}}.$$

3. Fractional-Order Bessel Functions

In this section, definitions of Bessel polynomials as well as their generalized fractional-order version are introduced. Hence, some properties and convergence results for them are established.

3.1. Bessel Polynomials

The Bessel polynomial $\mathbb{B}_n(x)$ of degree n and with constant term equal to 1 satisfies the following differential equation

$$x^2 y''(x) + 2(x+1) y'(x) - n(n+1) y(x) = 0, \quad n = 0, 1, \ldots.$$

Starting with $\mathbb{B}_0(x) = 1$ and $\mathbb{B}_1(x) = 1 + x$, the three-terms recurrence relation for the Bessel Polynomial is

$$\mathbb{B}_{n+1}(x) = (2n+1)x\,\mathbb{B}_n(x) + \mathbb{B}_{n-1}(x), \quad n = 1, 2, \ldots. \tag{7}$$

Beside $\mathbb{B}_0(x)$ and $\mathbb{B}_1(x)$, the next four of these polynomials are listed as follows

$$\begin{aligned}
\mathbb{B}_2(x) &= 1 + 3x + 3x^2, \\
\mathbb{B}_3(x) &= 1 + 6x + 15x^2 + 15x^3, \\
\mathbb{B}_4(x) &= 1 + 10x + 45x^2 + 105x^3 + 105x^4, \\
\mathbb{B}_5(x) &= 1 + 15x + 105x^2 + 420x^3 + 945x^4 + 945x^5.
\end{aligned}$$

The coefficients of these polynomials are positive with $\mathbb{B}_n(0) = 1$ and $\mathbb{B}'_n(0) = n(n+1)/2$. The explicit expression for the Bessel polynomials as the unique solution of the given differential equation is defined by

$$\mathbb{B}_n(x) = \sum_{k=0}^{n} \frac{1}{k!} \frac{(n+k)!}{(n-k)!} \left(\frac{x}{2}\right)^k, \quad n = 0, 1, \ldots. \tag{8}$$

These polynomials form an orthogonal system with respect to the weight function $w(x) \equiv \exp(-2/x)$ on the unite circle C, i.e.,

$$\frac{1}{2\pi i} \int_C \mathbb{B}_n(x)\,\mathbb{B}_m(x)\,w(x)dx = \frac{2(-1)^{n+1}\delta_{nm}}{2n+1}, \tag{9}$$

where δ_{nm} is the Kronecker delta function. Please note that the path of integration is not unique, and it can be replaced by an arbitrary curve surrounding $x = 0$. The same conclusion is true for the weight function $w(x)$. This implies that an arbitrary analytic function may be added to $w(x)$ and $w(x)$ may be multiplied by a nonzero constant. By means of the orthogonality relation (9), one may expand a function $g(x)$ in terms of Bessel functions

$$g(x) \approx \sum_{n=0}^{\infty} a_n\,\mathbb{B}_n(x),$$

where the coefficients a_n are

$$a_n = (-1)^{n+1}(n + \frac{1}{2}) \int_C \mathbb{B}_n(x)\,g(x)\,w(x)dx.$$

3.2. Fractional Bessel Polynomials

The fractional-order Bessel functions can be defined by introducing the change of variable $x = t^\alpha/L$, $L, \alpha > 0$ in (8). Let these polynomials will be denoted by $\mathbb{B}_n^\alpha(t) = \mathbb{B}_n(x)$. By generalizing these polynomials on the interval $[0, L]$ we obtain

$$\mathbb{B}_n^\alpha(t) = \sum_{k=0}^{n} \frac{\eta^k}{k!} \frac{(n+k)!}{(n-k)!} t^{k\alpha}, \quad 0 \leqslant t \leqslant L < \infty, \tag{10}$$

where $\eta = \frac{1}{2L}$. It is not difficult to show that the set of fractional polynomial functions $\{\mathbb{B}_0^\alpha, \mathbb{B}_1^\alpha, \ldots\}$ is orthogonal on $[0, L]$ with respect to the weight function $w_L^\alpha(t) \equiv t^{\alpha-1}\exp(-2L/t^\alpha)$. The fractional-order polynomials are useful in particular when the solutions of the underlying MOFDEs have fractional behavior.

3.3. Function Approximation and Convergence

Our goal is to obtain an approximate solution for the model problem (1) represented by the truncated Bessel series

$$x_{N,\alpha}(t) = \sum_{n=0}^{N} a_n \, \mathbb{B}_n^{\alpha}(t), \quad 0 \leqslant t \leqslant L, \tag{11}$$

where the unknown coefficients a_n, $n = 0, 1, \ldots, N$ must be sought. For this purpose, we express $\mathbb{B}_n^{\alpha}(t)$ in the matrix representation as

$$\boldsymbol{B}_{\alpha}(t) = \boldsymbol{T}_{\alpha}(t) \, \boldsymbol{D}^t, \tag{12}$$

where

$$\boldsymbol{T}_{\alpha}(t) = \begin{bmatrix} 1 & t^{\alpha} & t^{2\alpha} & \ldots & t^{N\alpha} \end{bmatrix}, \quad \boldsymbol{B}_{\alpha}(t) = [\mathbb{B}_0^{\alpha}(t) \quad \mathbb{B}_1^{\alpha}(t) \quad \ldots \quad \mathbb{B}_N^{\alpha}(t)],$$

and the lower triangular matrix $\boldsymbol{D}$ of size $(N+1) \times (N+1)$ takes the form

$$\boldsymbol{D} = \begin{bmatrix} 1 & 0 & 0 & \ldots & 0 & 0 \\ 1 & 1 & 0 & \ldots & 0 & 0 \\ 1 & 3 & 3 & \ldots & 0 & 0 \\ \vdots & \vdots & \ddots & \ddots & \ddots & \vdots \\ 1 & \dfrac{\eta \, N!}{(N-2)! \, 1!} & \dfrac{\eta^2 \, (N+1)!}{(N-3)! \, 2!} & \ldots & \dfrac{\eta^{N-1} \, (2N-2)!}{0! \, (N-1)!} & 0 \\ 1 & \dfrac{\eta \, (N+1)!}{(N-1)! \, 1!} & \dfrac{\eta^2 \, (N+2)!}{(N-2)! \, 2!} & \ldots & \dfrac{\eta^{N-1} \, (2N-1)!}{1! \, (N-1)!} & \dfrac{\eta^N \, (2N)!}{0! \, N!} \end{bmatrix}.$$

By expressing the relation (11) in a matrix form and exploiting (12), the approximate solution $x_{N,\alpha}(t)$ in the matrix form can be rewritten as

$$x_{N,\alpha}(t) = \boldsymbol{B}_{\alpha}(t) \, \boldsymbol{A} = \boldsymbol{T}_{\alpha}(t) \, \boldsymbol{D}^t \, \boldsymbol{A}, \tag{13}$$

where the vector of unknown is $\boldsymbol{A} = [a_0 \quad a_1 \quad \ldots \quad a_N]^t$. Our further aim is to establish the convergence results of the fractional Bessel polynomials. Roughly speaking, the next theorem shows that the approximate solution $x_{N,\alpha}(t)$ converges to the solution $x(t)$ of differential Equation (1) as $N \to \infty$, see e.g., [21] for a similar proof.

Theorem 2. *Let assume that $\mathcal{D}_{\star}^{k\alpha} g(t) \in C(0, L]$ for $k = 0, 1, \ldots, N$ and let*

$$\mathcal{S}_N^{\alpha} = Span\langle \mathbb{B}_0^{\alpha}(t), \mathbb{B}_1^{\alpha}(t), \ldots, \mathbb{B}_{N-1}^{\alpha}(t) \rangle.$$

Suppose that $g_{N,\alpha}(t) = \boldsymbol{B}_{\alpha}(t) \, \boldsymbol{A}$ is the best approximation out of $\mathcal{S}_N^{\alpha}$ to g, then the following error bound holds:

$$\| g(t) - g_{N,\alpha}(t) \|_{w_L^{\alpha}} \leqslant \frac{L^{N\alpha} M^{\alpha}}{\exp(\frac{1}{L^{\alpha}-1}) \, \Gamma(N\alpha + 1)} \Big(\frac{L^{\alpha}}{2N\alpha + \alpha} \Big)^{1/2},$$

where $M^{\alpha} \geqslant |\mathcal{D}_{\star}^{N\alpha} g(t)|$, $t \in (0, L]$.

Proof. According to Theorem 1, the generalized Taylor's formula for $g(t)$ can be represented as $G = \sum_{j=0}^{N-1} \frac{t^{j\alpha}}{\Gamma(j\alpha+1)} \mathcal{D}_\star^{j\alpha} g(0^+)$, and satisfies

$$|g - G| \leqslant \frac{t^{N\alpha}}{\Gamma(N\alpha+1)} M^\alpha.$$

Using the fact that $\boldsymbol{B}_\alpha(t)\,\boldsymbol{A}$ is the best approximation to g from $\mathcal{S}_N^\alpha$ and $G \in \mathcal{S}_N^\alpha$, we conclude that

$$\|g(t) - g_{N,\alpha}(t)\|_{w_L^\alpha}^2 \leqslant \|g - G\|_{w_L^\alpha}^2 \leqslant \Big[\frac{M^\alpha}{\Gamma(N\alpha+1)}\Big]^2 \int_0^L \exp(-\frac{2L}{t^\alpha}) t^{2N\alpha} t^{\alpha-1} dt. \tag{14}$$

Employing the inequality $-\frac{2L^\alpha}{t^\alpha} \leqslant -2$, which holds for all $t \in (0, L]$, one immediately find that $\exp(-\frac{2L}{t^\alpha}) \leqslant \exp(\frac{-2}{L^{\alpha-1}})$. Thus, by inserting this inequality into (14) and then integrating we conclude that

$$\|g(t) - g_{N,\alpha}(t)\|_{w_L^\alpha}^2 \leqslant \Big[\frac{M^\alpha}{\Gamma(N\alpha+1)}\Big]^2 \frac{\exp(\frac{-2}{L^{\alpha-1}}) L^{(2N+1)\alpha}}{(2N+1)\alpha}.$$

The proof is complete by taking the square roots of both sides. □

Therefore, for obtaining an approximate solution of the form (11) for the solution of (1) the following collocation points are used on $0 < t \leqslant L$,

$$t_i = \frac{L}{N} i, \quad i = 0, 1, \ldots, N. \tag{15}$$

4. The Collocation Scheme

To proceed, we approximate the solution $x(t)$ of MOFDEs (1) in terms of $(N+1)$-terms Bessel polynomials series denoted by $x_{N,\alpha}(t)$ on the interval $[0, L]$. In the matrix representation, we consider

$$x(t) \cong x_{N,\alpha}(t) = \boldsymbol{T}_\alpha(t)\, \boldsymbol{D}^t\, \boldsymbol{A}. \tag{16}$$

By placing the collocation points (15) into (16), we get to a system of matrix equations as

$$x_{N,\alpha}(t_i) = \boldsymbol{T}_\alpha(t_i)\, \boldsymbol{D}^t\, \boldsymbol{A}, \quad i = 0, 1, \ldots, N.$$

Hence, we write the preceding equations compactly as

$$\boldsymbol{X} = \boldsymbol{T}\, \boldsymbol{D}^t\, \boldsymbol{A}, \tag{17}$$

where

$$\boldsymbol{T} = \begin{bmatrix} \boldsymbol{T}_\alpha(t_0) \\ \boldsymbol{T}_\alpha(t_1) \\ \vdots \\ \boldsymbol{T}_\alpha(t_N) \end{bmatrix}, \quad \boldsymbol{X} = \begin{bmatrix} x_{N,\alpha}(t_0) \\ x_{N,\alpha}(t_1) \\ \vdots \\ x_{N,\alpha}(t_N) \end{bmatrix}.$$

To handle the fractional derivative of order γ in (1), we differentiate both sides of (16),

$$\mathcal{D}_\star^\gamma x_{N,\alpha}(t) = \mathcal{D}_\star^\gamma\, \boldsymbol{T}_\alpha(t)\, \boldsymbol{D}^t\, \boldsymbol{A}. \tag{18}$$

By means of the property (5) and (6), the calculation of $\mathcal{D}_\star^\gamma\, \boldsymbol{T}_\alpha(t)$ can be easily obtained as follows

$$\boldsymbol{T}_\alpha^{(\gamma)}(t) := \mathcal{D}_\star^\gamma\, \boldsymbol{T}_\alpha(t) = [0 \quad \mathcal{D}_\star^\gamma t^\alpha \quad \ldots \quad \mathcal{D}_\star^\gamma t^{\alpha N}].$$

To write the fractional derivative $\mathcal{D}_\star^\gamma$ involved in (1) in the matrix form, the collocation points (15) will be inserted into (18) to have

$$\mathcal{D}_\star^\gamma x_{N,\alpha}(t_i) = \boldsymbol{T}_\alpha^{(\gamma)}(t_i)\, \mathbf{D}^t \boldsymbol{A}, \quad i = 0, 1 \ldots, N,$$

which can be expressed equivalently as

$$\mathbf{X}^{(\gamma)} = \boldsymbol{T}^{(\gamma)}\, \mathbf{D}^t \boldsymbol{A}, \tag{19}$$

where

$$\mathbf{X}^{(\gamma)} = \begin{bmatrix} \mathcal{D}_\star^\gamma x_{N,\alpha}(t_0) \\ \mathcal{D}_\star^\gamma x_{N,\alpha}(t_1) \\ \vdots \\ \mathcal{D}_\star^\gamma x_{N,\alpha}(t_N) \end{bmatrix}, \quad \boldsymbol{T}^{(\gamma)} = \begin{bmatrix} \boldsymbol{T}_\alpha^{(\gamma)}(t_0) \\ \boldsymbol{T}_\alpha^{(\gamma)}(t_1) \\ \vdots \\ \boldsymbol{T}_\alpha^{(\gamma)}(t_N) \end{bmatrix}.$$

Similarly, the fractional derivative operators $\mathcal{D}_\star^{\beta_j} x(t)$ in (1) for $j = 1, \ldots, \ell$ can be approximated as

$$\mathbf{X}^{(\beta_j)} = \boldsymbol{T}^{(\beta_j)}\, \mathbf{D}^t \boldsymbol{A}, \tag{20}$$

where $\mathbf{X}^{(\beta_j)}$ and $\boldsymbol{T}^{(\beta_j)}$ are obtained as in (20) by replacing γ with β_j.

By inserting the collocation points into (1), we have the system

$$\mathcal{D}_\star^\gamma x(t_i) = F\left(t_i, x(t_i), \mathcal{D}_\star^{\beta_1} x(t_i), \ldots, \mathcal{D}_\star^{\beta_\ell} x(t_i)\right), \quad i = 0, 1, \ldots, N. \tag{21}$$

Considering these equations in a matrix form and substituting the relations (17), (19), and (20) into the resulting system, a fundamental matrix equation is obtained to be solved. Let us assume that the function F in (21) is the linear form

$$F = \sum_{k=1}^{\ell} c_k(t)\, \mathcal{D}_\star^{\beta_k} x(t) + c_0(t)\, x(t) + h(t),$$

where $c_k(t)$ for $k = 1, \ldots, \ell$ and $c_0(t)$, $h(t)$ are given functions. In this case, the equations in (21) can be rewritten in the matrix representation as

$$\mathbf{X}^{(\gamma)} = \sum_{k=1}^{\ell} \mathbf{C}_k\, \mathbf{X}^{(\beta_k)} + \mathbf{C}_0\, \mathbf{X} + \boldsymbol{H}, \tag{22}$$

where the coefficient matrices $\mathbf{C}_k$, $k = 0, 1, \ldots \ell$ with size $(N+1) \times (N+1)$ and the vector $\boldsymbol{H}$ of size $(N+1) \times 1$ have the forms

$$\boldsymbol{C}_k = \begin{bmatrix} c_k(t_0) & 0 & \ldots & 0 \\ 0 & c_k(t_1) & \ldots & 0 \\ \vdots & \vdots & \ddots & \vdots \\ 0 & 0 & \ldots & c_k(t_N) \end{bmatrix}, \quad \boldsymbol{H} = \begin{bmatrix} h(t_0) \\ h(t_1) \\ \vdots \\ h(t_N) \end{bmatrix},$$

Substituting the relations (17), (19), and (20) into (22), the fundamental matrix equation is obtained

$$\boldsymbol{W}\boldsymbol{A} = \boldsymbol{H}, \quad \text{or} \quad [\boldsymbol{W}; \boldsymbol{H}], \tag{23}$$

where

$$\boldsymbol{W} := \left(\boldsymbol{T}^{(\gamma)} - \mathbf{C}_0\, \boldsymbol{T} - \mathbf{C}_1\, \boldsymbol{T}^{(\beta_1)} - \ldots - \mathbf{C}_\ell\, \boldsymbol{T}^{(\beta_\ell)}\right) \mathbf{D}^t.$$

Obviously, (23) is a linear matrix equation and a_n, $n = 0, 1, \ldots, N$ are the unknowns Bessel coefficients to be determined.

Next aim is to take into account the initial or boundary conditions (2). For the first condition $x(0) = x_0$, we tend $t \to 0$ in (16) to get the following matrix representation

$$\widehat{\boldsymbol{X}}_0 \boldsymbol{A} = x_0, \qquad \widehat{\boldsymbol{X}}_0 := \boldsymbol{T}_\alpha(0)\, \boldsymbol{D}^t = [\hat{x}_{00} \quad \hat{x}_{01} \quad \ldots \quad \hat{x}_{0N}].$$

For the remaining initial conditions, one needs to calculate the integer-order derivatives $\frac{d^k}{dt^k}\boldsymbol{T}_\alpha(t)$, $k = 1, 2, \ldots, n-1$, which strictly depend on α as well as N. For example, by choosing $\alpha = 1/2$ and $N = 7$ we get

$$\boldsymbol{T}_{\frac{1}{2}}(t) = \begin{bmatrix} 1 & t^{1/2} & t & t^{3/2} & t^2 & t^{5/2} & t^3 & t^{7/2} \end{bmatrix}.$$

Differentiation twice with respect to t reveals that

$$\begin{aligned} \frac{d}{dt}\boldsymbol{T}_{\frac{1}{2}}(t) &= \begin{bmatrix} 0 & 0 & 1 & \frac{3}{2}t^{1/2} & 2t & \frac{5}{2}t^{3/2} & 3t^2 & \frac{7}{2}t^{5/2} \end{bmatrix}, \\ \frac{d^2}{dt^2}\boldsymbol{T}_{\frac{1}{2}}(t) &= \begin{bmatrix} 0 & 0 & 0 & 0 & 2 & \frac{15}{4}t^{1/2} & 6t & \frac{35}{4}t^{3/2} \end{bmatrix}. \end{aligned}$$

Now, by differentiating k times in (16), and defining

$$\boldsymbol{T}_\alpha^{(k)}(t) := \frac{d^k}{dt^k}\boldsymbol{T}_\alpha(t),$$

with the limit $t \to 0$, we conclude for $k = 1, 2, \ldots, n-1$ that

$$\widehat{\boldsymbol{X}}_k \boldsymbol{A} = x_k, \qquad \widehat{\boldsymbol{X}}_k := \boldsymbol{T}_\alpha^{(k)}(0)\, \boldsymbol{D}^t = [\hat{x}_{k0} \quad \hat{x}_{k1} \quad \ldots \quad \hat{x}_{kN}].$$

Similarly, for the end conditions $x^{(k)}(L) = x_{Lk}$, $k = 0, \ldots, n-1$, the following matrix expressions are obtained

$$\widehat{\boldsymbol{X}}_{Lk} \boldsymbol{A} = x_{Tk}, \qquad \widehat{\boldsymbol{X}}_{Lk} := \boldsymbol{T}_\alpha^{(k)}(L)\, \boldsymbol{D}^t = [\hat{x}_{L0} \quad \hat{x}_{L1} \quad \ldots \quad \hat{x}_{LN}].$$

Now, we replace the first n rows of the augmented matrix $[\boldsymbol{W}; \boldsymbol{H}]$ in (23) by the row matrices $[\widehat{\boldsymbol{X}}_k; x_k]$ or $[\widehat{\boldsymbol{X}}_{Lk}; x_{Lk}]$, $k = 0, 1, \ldots, n-1$ to get the (nonlinear) algebraic system of equations

$$\widehat{\boldsymbol{W}}\boldsymbol{A} = \widehat{\boldsymbol{H}}, \quad \text{or} \quad [\widehat{\boldsymbol{W}}; \widehat{\boldsymbol{H}}].$$

Thus, the unknown Bessel coefficients in (16) will be known through solving this (nonlinear) system. This can be obtained by using the Newton's iterative algorithm.

Remark 1. *In numerical applications below, we frequently encounter the nonlinear terms like $x^s(t)$ for $s = 2, 3 \ldots$. To approximate the nonlinear term $x^2(t)$ in terms of $x^2_{N,\alpha}(t)$, the collocation points (15) will be substituted into $x^2_{N,\alpha}(t)$. It can be easily seen that in the matrix representation we have*

$$\boldsymbol{X}^2 = \begin{bmatrix} x^2_{N,\alpha}(t_0) \\ x^2_{N,\alpha}(t_1) \\ \vdots \\ x^2_{N,\alpha}(t_N) \end{bmatrix} = \begin{bmatrix} x_{N,\alpha}(t_0) & 0 & \ldots & 0 \\ 0 & x_{N,\alpha}(t_1) & \ldots & 0 \\ \vdots & \vdots & \ddots & \vdots \\ 0 & 0 & \ldots & x_{N,\alpha}(t_N) \end{bmatrix} \begin{bmatrix} x_{N,\alpha}(t_0) \\ x_{N,\alpha}(t_1) \\ \vdots \\ x_{N,\alpha}(t_N) \end{bmatrix} = \widehat{\boldsymbol{X}}\boldsymbol{X}.$$

Using (16), we further express the matrix $\widehat{\boldsymbol{X}}$ as a product of three block diagonal matrices as follows

$$\widehat{\boldsymbol{X}} = \widehat{\boldsymbol{T}}\,\widehat{\boldsymbol{D}}\,\widehat{\boldsymbol{A}},$$

where

$$\widehat{T} = \begin{bmatrix} T_\alpha(t_0) & 0 & \dots & 0 \\ 0 & T_\alpha(t_1) & \dots & 0 \\ \vdots & \vdots & \ddots & \vdots \\ 0 & 0 & \dots & T_\alpha(t_N) \end{bmatrix}, \widehat{D} = \begin{bmatrix} D^t & 0 & \dots & 0 \\ 0 & D^t & \dots & 0 \\ \vdots & \vdots & \ddots & \vdots \\ 0 & 0 & \dots & D^t \end{bmatrix}, \widehat{A} = \begin{bmatrix} A & 0 & \dots & 0 \\ 0 & A & \dots & 0 \\ \vdots & \vdots & \ddots & \vdots \\ 0 & 0 & \dots & A \end{bmatrix}.$$

Analogously, the higher-order nonlinear terms can be treated recursively $\mathbf{X}^s = (\widehat{\mathbf{X}})^{s-1} X$, $s = 3, 4, \ldots$.

4.1. Error Estimation

In general, the exact solution of most MOFDEs cannot be explicitly obtained. Thus, we need some measurements to test the accuracy of the proposed scheme. Since the truncated Bessel series (11) as an approximate solution is satisfied in (1), our expectation is that the residual error function denoted by $\mathcal{R}_{N,\alpha}(t)$ becomes approximately small. Here, $\mathcal{R}_{N,\alpha}(t) : [0, L] \to \mathbb{R}$ obtained by inserting the computed approximated solution $x_{N,\alpha}(t)$ into the differential equation (1). More precisely, for testing accuracy of some numerical models we calculate

$$\mathcal{R}_{N,\alpha}(t) = \left| \mathcal{D}_\star^\gamma x_{N,\alpha}(t) - F\left(t, x_{N,\alpha}(t), \mathcal{D}_\star^{\beta_1} x_{N,\alpha}(t), \ldots, \mathcal{D}_\star^{\beta_\ell} x_{N,\alpha}(t)\right) \right| \cong 0, \quad t \in [0, L]. \tag{24}$$

It should be noticed that the fractional derivatives of order γ, β_j, $j = 1, \ldots, \ell$ of the approximate solution $x_{N,\alpha}(t)$ in (24) are calculated by using the properties (5) and (6). Obviously, the residual function is vanished at the collocation points (15), so our expectation is that $\mathcal{R}_{N,\alpha}(t) \to 0$ as N tends to infinity. This implies that the smallness of the residual error function shows the closeness of the approximate solution to the true exact solution.

5. Illustrative Test Problems

Now, we show the benefits of the presented Bessel-collocation scheme by simulating some case examples including various linear and nonlinear initial and boundary value problems. The numerical models and calculations are verified through a comparison with existing computational schemes and experimental measurements. Our computations were carried out using MATLAB software version R2017a.

Problem 1. *In the first problem, we consider the following inhomogeneous Bagley–Torvik equation modelling the motion of an immersed plate in a Newtonian fluid [5–7]*

$$x^{(2)}(t) + \mathcal{D}_\star^{\frac{3}{2}} x(t) + x(t) = t + 1,$$

with the initial conditions $x(0) = 1$ *and* $x^{(1)}(0) = 1$. *The exact solution of this problem is* $x(t) = t + 1$.

By employing $N = 2$ and $L = 1$, we are looking for an approximate solution in the form $x_{N,\alpha}(t) = \sum_{n=0}^{2} a_n \mathbb{B}_n^\alpha(t)$. To this end, we calculate the unknown coefficients a_0, a_1, and a_2. For this example we set $\alpha = 1$ and the collocation points $t_0 = 0$, $t_1 = \frac{1}{2}$, $t_3 = 0$ are used. Using $\gamma = 2$ and $\beta_1 = \frac{3}{2}$, the corresponding matrices and vectors in the fundamental matrix Equation (23) become

$$T^{(\frac{3}{2})} = \begin{bmatrix} 0 & 0 & 0 \\ 0 & 0 & 1358/851 \\ 0 & 0 & 167/74 \end{bmatrix}, \quad T^{(2)} = \begin{bmatrix} 0 & 0 & 2 \\ 0 & 0 & 2 \\ 0 & 0 & 2 \end{bmatrix}, \quad T = \begin{bmatrix} 1 & 0 & 0 \\ 1 & 1/2 & 1/4 \\ 1 & 1 & 1 \end{bmatrix},$$

$$D = \begin{bmatrix} 1 & 0 & 0 \\ 1 & 1 & 0 \\ 1 & 3 & 3 \end{bmatrix}, \quad H = \begin{bmatrix} 1 \\ 3/2 \\ 2 \end{bmatrix}, \quad \left[\widehat{W};\widehat{H}\right] = \begin{bmatrix} 1 & 1 & 1 & ; & 1 \\ 1 & 3/2 & 1881/134 & ; & 3/2 \\ 0 & 1 & 3 & ; & 1 \end{bmatrix}.$$

By solving the linear system $\widehat{W}A = \widehat{H}$, the coefficients matrix is found as

$$A = [0 \quad 1 \quad 0]^t.$$

Afterwards, by inserting the obtained coefficients into $x_{2,1}(t)$ we get the approximate solution

$$x_{2,1}(t) = \begin{bmatrix} 1 & 1+t & 3t^2+3t+1 \end{bmatrix} A = 1+t,$$

which is the desired exact solution.

Problem 2. *In the next example, the following nonlinear initial-value problem will be considered [5,7]*

$$x^{(3)}(t) + \mathcal{D}_{\star}^{\frac{5}{2}} x(t) + x^2(t) = t^4.$$

The initial conditions are $x(0) = 0$, $x^{(1)}(0) = 0$, and $x^{(2)}(0) = 2$. It can be easily checked that the exact true solution is $x(t) = t^2$.

For this example, we take $N = 3$, $\alpha = 1$, and the collocation points are $\{0, \frac{1}{3}, \frac{2}{3}, 1\}$. To obtain the unknown coefficients a_0, a_1, a_2, a_3 in $x_{3,1}(t)$, the following nonlinear algebraic system of equations to be solved

$$\begin{cases} a_0 + a_1 + a_2 + a_3 = 0, \\ a_1 + 3a_2 + 6a_3 = 0, \\ 6a_2 + 30a_3 = 2, \\ 90a_3 + \frac{180}{\sqrt{\pi}} a_3 + (a_0 + 2a_1 + 7a_2 + 37a_3)^2 = 1. \end{cases}$$

By solving the above system, we get

$$a_0 = \frac{2}{3}, \quad a_1 = -1, \quad a_2 = \frac{1}{3}, \quad a_3 = 0.$$

Therefore, we get

$$x_{3,1}(t) = \begin{bmatrix} 1 & 1+t & 1+3t+3t^2 & 1+6t+15t^2+15t^3 \end{bmatrix} A = t^2,$$

which is obviously the exact solution.

In the next example, we show the advantage of using the fractional-order Bessel functions in the computations.

Problem 3. *In this test example, we solve the initial-value problem [7]*

$$x^{(1)}(t) + \mathcal{D}_{\star}^{\frac{1}{2}} x(t) + x(t) = t^{\frac{5}{2}} + \frac{5}{2} t^{\frac{3}{2}} + \frac{15\sqrt{\pi}}{16} t^2,$$

with initial condition $x(0) = 0$. The exact solution is $x(t) = t^2\sqrt{t}$.

We first consider $N = 5$ and $\alpha = 1/2$. The approximated solution $x_{5,\frac{1}{2}}(t)$ for $t \in [0,1]$ takes the form

$$x_{5,\frac{1}{2}}(t) = 3.55309 \times 10^{-14}\, t^2 - 2.20767 \times 10^{-13}\, t + 2.38332 \times 10^{-13}\, t^{1/2} + 4.17169 \times 10^{-14}\, t^{3/2} + 1.0\, t^{5/2} + 9.97959 \times 10^{-111}.$$

However, with a lower number of basis functions one can also obtain an accurate result. Using $N = 2$, $\alpha = 5/2$ and $N = 3$, $\alpha = 5/6$, the following approximations are obtained

$$x_{2,\frac{5}{2}}(t) = 4.88118 \times 10^{-17}\, t^5 + 1.0\, t^{5/2},$$
$$x_{3,\frac{5}{6}}(t) = 1.0\, t^{5/2} - 4.899133356 \times 10^{-17}\, t^{5/3} + 5.889017161 \times 10^{-17}\, t^{5/6}.$$

Moreover, to show the advantage of the presented approach and to validate our obtained approximated solutions, we make a comparison in terms of errors in the L_∞ and L_2 norms in Table 1. We compare the Bessel-collocation approach and the Chelyshkov collocation spectral method [7]. In this comparison, we use different $N = 1,2,3$ and $\alpha = 1/2, 5/2, 5/6$.

Table 1. Comparison of L_∞, L_2 error norms for test Problem 3.

	New Bessel			Chelyshkov [7]	
	$N = 2, \alpha = \frac{5}{2}$	$N = 3, \alpha = \frac{5}{6}$	$N = 5, \alpha = \frac{1}{2}$	$N = 16$	$N = 20$
L_∞	3.17_{-17}	2.50_{-17}	8.05_{-14}	2.45_{-06}	8.59_{-07}
L_2	6.18_{-17}	6.32_{-17}	2.39_{-14}	9.89_{-07}	3.24_{-07}

Problem 4. *Consider the boundary value problem [22,23]*

$$\mathcal{D}_\star^\gamma x(t) - \mathcal{D}_\star^\beta x(t) = -(1 + \exp(t-1)), \quad 1 < \gamma \leqslant 2, \quad 0 < \beta \leqslant 1,$$

with initial conditions $x(0) = 0$ and $x(1) = 0$. The exact solution corresponds to $\gamma = 2$ and $\beta = 1$ is given as $x(t) = t - t\exp(t-1)$.

Let $N = 8$ and set $\alpha = 1$. For $\gamma = 2$, $\beta = 1$, the approximate solution $x_{8,1}(t)$ of the model Problem 4 using Bessel functions in the interval $0 \leqslant t \leqslant 1$ is

$$x_{8,1}(t) = -0.0001286702494\, t^8 - 0.0003938666636\, t^7 - 0.003196278513\, t^6 - 0.01524130813\, t^5 - 0.06134909192\, t^4 - 0.183930672\, t^3 - 0.3678807668\, t^2 + 0.6321206543\, t - 2.12897992 \times 10^{-109}.$$

In Table 2, we report the numerical results corresponding to these values of γ, β using different $N = 8, 16$ evaluated at some points $t \in [0,1]$. The corresponding absolute errors $\mathcal{E}_{N,\alpha}(t) := |x(t) - x_{N,\alpha}(t)|$ are also reported in this table. Moreover, the numerical results based on Haar wavelet operational matrices [22] are given in the last column of Table 2. As can see from Table 2, our approximate solutions agree with the results obtained in [22]. The next observation is that more accurate solutions are obtained if one increases the number of Bessel functions N.

Table 2. Comparison of approximate solutions and the corresponding absolute errors for test Problem 4 using $N = 8, 16$, and $\gamma = 2$, $\beta = 1$.

	$N = 8$		$N = 16$		**Haar Wavelet** [22]
t	$x_{8,1}(t)$	$\mathcal{E}_{8,1}(t)$	$x_{16,1}(t)$	$\mathcal{E}_{16,1}(t)$	$J = 10$
0.1	0.0593430365264982	2.50_{-09}	0.059343034025940	2.96_{-16}	0.05934300
0.2	0.1101342091046447	1.93_{-09}	0.110134207176555	3.02_{-16}	0.11013418
0.3	0.1510244104291788	1.57_{-09}	0.151024408862577	3.07_{-16}	0.15102438
0.4	0.1804753466639111	1.10_{-09}	0.180475345562389	2.63_{-16}	0.18047531
0.5	0.1967346707501356	6.06_{-10}	0.196734670143683	2.57_{-16}	0.19673463
0.6	0.1978079724388553	6.02_{-11}	0.197807972378616	2.13_{-16}	0.19780792
0.7	0.1814272449585667	5.64_{-10}	0.181427245522797	2.80_{-16}	0.18142718
0.8	0.1450153963505055	1.19_{-09}	0.145015397537614	1.73_{-16}	0.14501532
0.9	0.0856463216234075	2.14_{-09}	0.085646323767636	1.52_{-16}	0.08564623

In Figure 1, $x_{10,1}(t)$ is plotted when $\gamma = 2$ ($\beta = 1$) is fixed and different values of $\beta = 0.25, 0.5, 0.75, 1$ ($\gamma = 1.25, 1.5, 1.75, 2$) are examined. It is observed that as γ and β approached to 1 and 2 respectively, numerical solutions tend to the exact solutions.

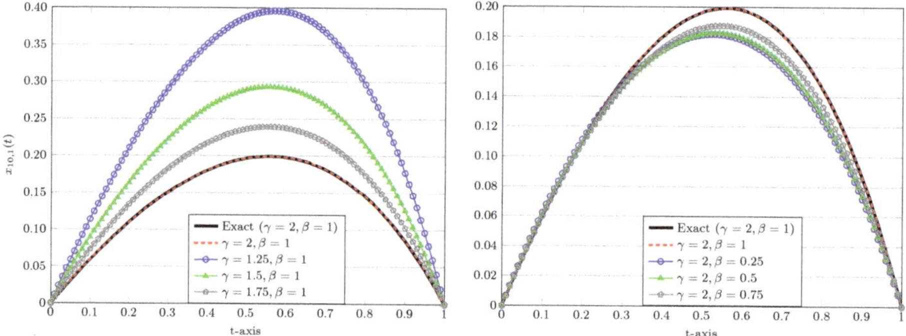

Figure 1. Numerical approximations for fixed $\beta = 1$ and $\gamma = 1.25, 1.5, 1.75, 2$ (**left**) and fixed $\gamma = 2$ and $\beta = 0.25, 0.5, 0.75, 1$ (**right**) in test Problem 4 with $N = 10$.

Problem 5. *Let us consider the initial-value problem of Bagley–Torvik equation of fractional order with variable coefficients [24,25]*

$$x^{(2)}(t) + \frac{1}{2}\sqrt{\pi}t^2\, \mathcal{D}_\star^{\frac{3}{2}} x(t) - 4\sqrt{t}\, x(t) = 6t,$$

with initial conditions $x(0) = 0$, $x^{(1)}(0) = 0$. *The exact solution is* $x(t) = t^3$.

Clearly, the exact solution is a third-degree polynomial. Therefore, we take $N = 3$ and $\alpha = 1$, which are sufficient to get the desired approximations. Using the usual collocation points as in Problem 2 and similar to Problem 1, we get the final augmented matrix

$$\left[\widehat{\mathbf{W}}; \widehat{\boldsymbol{H}}\right] = \begin{bmatrix} 1 & 1 & 1 & 1 & ; & 0 \\ -1351/585 & -4782/1553 & 2691/2701 & 6598/129 & ; & 2 \\ -1277/391 & -4801/882 & -2659/445 & 6090/97 & ; & 4 \\ 0 & 1 & 3 & 6 & ; & 0 \end{bmatrix}.$$

Solving the resulting linear system, we find

$$\boldsymbol{A} = [-\frac{1}{3} \quad \frac{3}{5} \quad -\frac{1}{3} \quad \frac{1}{15}]^t.$$

Therefore, the approximated solution $x_{3,1}(t)$ is obtained as

$$x_{3,1}(t) = \begin{bmatrix} 1 & 1+t & 3t^2+3t+1 & 1+6t+15t^2+15t^3 \end{bmatrix} \boldsymbol{A} = t^3,$$

which is the exact solution.

Problem 6. *Consider the fractional Riccati equation [23,26]*

$$\mathcal{D}_\star^\gamma x(t) + x(t) - x^2(t) = 0, \quad 0 < \gamma \leqslant 1,$$

on a long time interval with $L = 5$ *and initial condition* $x(0) = 1/2$. *When* $\gamma = 1$, *the exact solution is* $x(t) = \frac{1}{\exp(t)+1}$.

We calculate the approximated solution $x_{N,\alpha}(t)$ using $N = 7$ and γ equals to $\alpha = 1/4$. Thus, we get

$$\begin{aligned} x_{7,\frac{1}{4}}(t) &= 0.1617950181136742\, t^{\frac{3}{4}} - 0.03445544072182753\, t^{\frac{1}{2}} - 0.2700823207417999\, t^{\frac{1}{4}} \\ &- 0.009800299427063008\, t^{\frac{3}{2}} - 0.1204495580962043\, t + 0.04675854257899483\, t^{\frac{5}{4}} \\ &+ 0.0008798927221707359\, t^{\frac{7}{4}} + 0.49999999999998357401. \end{aligned}$$

To validate this solution, we also employ the old fractional-order Bessel polynomials as well as Chelyshkov and Legendre functions from the previous works [26,27] with the same parameters as above. The corresponding solutions take the forms respectively

$$\begin{aligned} x^B_{7,\frac{1}{4}}(t) &= 0.1617932518503192\, t^{\frac{3}{4}} - 0.03445464899775196\, t^{\frac{1}{2}} - 0.27008246876491873\, t^{\frac{1}{4}} \\ &- 0.0097998224671427077\, t^{\frac{3}{2}} - 0.1204474711992175\, t + 0.04675716925988139\, t^{\frac{5}{4}} \\ &+ 0.00087982440038136711651\, t^{\frac{7}{4}} + 0.5, \end{aligned}$$

$$\begin{aligned} x^C_{7,\frac{1}{4}}(t) &= 0.16176395176134591\, t^{\frac{3}{4}} - 0.034436828673633131\, t^{\frac{1}{2}} - 0.27008706550355819\, t^{\frac{1}{4}} \\ &- 0.0097962860256647\, t^{\frac{3}{2}} - 0.12042146216589336\, t + 0.046744073064058187\, t^{\frac{5}{4}} \\ &+ 0.00087942518183550405624\, t^{\frac{7}{4}} + 0.5, \end{aligned}$$

$$\begin{aligned} x^L_{7,\frac{1}{4}}(t) &= 0.16179490530760574\, t^{\frac{3}{4}} - 0.034455365809483707\, t^{\frac{1}{2}} - 0.2700823428770175\, t^{\frac{1}{4}} \\ &- 0.009800287282136090\, t^{\frac{3}{2}} - 0.12044946374788617\, t + 0.04675849679405389\, t^{\frac{5}{4}} \\ &+ 0.00087989135193320893222\, t^{\frac{7}{4}} + 0.4999999947316410565 \end{aligned}$$

To further compare these collocation schemes based on various polynomials, we calculate the estimated residual errors obtained by the relation (24). The graphs of $\mathcal{R}_{N,\alpha}(t)$ on the interval $[0,5]$ correspond to $\gamma, \alpha = 1/4$ and for $N = 7$ are shown in Figure 2. With respect to Figure 2, it is obviously seen that the residual error functions obtained by the presented Bessel-collocation method are smaller compared to the errors of other polynomial-based numerical collocation schemes.

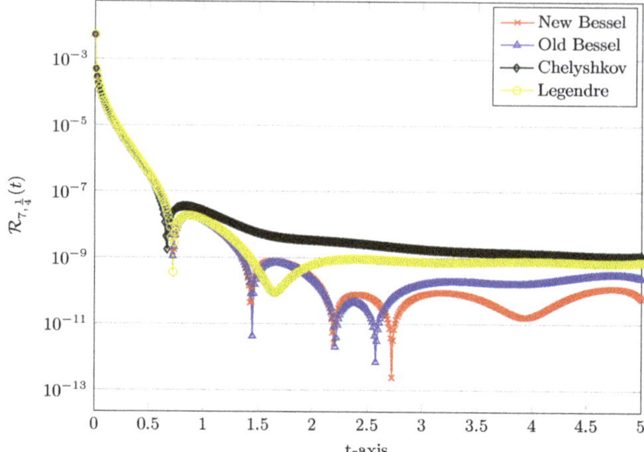

Figure 2. Comparing the error functions in test Problem 6 using old and new Bessel, Chelyshkov, and Legendre functions with $\gamma, \alpha = 1/4$, and $N = 7$.

Problem 7. *Consider the following nonlinear boundary value problem with variable coefficients [6]*

$$x^{(2)}(t) + \Gamma(\frac{4}{5})\sqrt[5]{t^6}\,\mathcal{D}_\star^{\frac{6}{5}} x(t) + \frac{11}{9}\Gamma(\frac{5}{6})\sqrt[6]{t}\,\mathcal{D}_\star^{\frac{1}{6}} x(t) - [x^{(1)}(t)]^2 = 2 + \frac{1}{10}t^2, \quad 0 < t < 1,$$

with boundary conditions $x(0) = 1$ and $x(1) = 2$. The exact solution of this example is $x(t) = 1 + t^2$.

In this example, we have $\gamma = 2$, $\beta_1 = 6/5$, and $\beta_2 = 1/6$. First, we set $\alpha = 1$. The approximate solutions $x_{N,\alpha}(t)$ of Problem 7 for $N = 2, 3$ on $0 \leqslant t \leqslant 1$ are obtained as follows, respectively

$$\begin{aligned} x_{2,1}(t) &= 1.0000000000119503851\, t^2 - 2.98118 \times 10^{-11}\, t + 1.0000000000072342687, \\ x_{3,1}(t) &= 1.12820 \times 10^{-9}\, t^3 + 1.0000000012230527969\, t^2 - 9.66432 \times 10^{-11}\, t \\ &\quad + 1.0000000000012118324. \end{aligned}$$

To show the gain of the proposed scheme, we compare our results with the collocation method based on Bernstein operational matrix (BOM) of fractional derivative from [6]. Table 3 reports the errors in L_∞ and L_2 norms of the new Bessel-collocation procedure and the errors of the BOM algorithm. This comparison shows the thoroughness of the proposed method.

Table 3. Comparison of L_∞, L_2 error norms for test Problem 7.

	New Bessel		BOM [6]			
	$N = 2$	$N = 3$	$N = 3$	$N = 6$	$N = 12$	$N = 15$
L_∞	1.06271_{-11}	1.45886_{-10}	3.4_{-05}	1.5_{-06}	5.5_{-08}	1.9_{-08}
L_2	3.00764_{-11}	4.21226_{-10}	2.0_{-05}	7.6_{-07}	2.3_{-08}	7.9_{-09}

Problem 8. *We consider the following initial-value problem of multi-term nonlinear fractional differential equation [6]*

$$\mathcal{D}_\star^{\gamma} x(t) + \mathcal{D}_\star^{\beta_1} x(t) \cdot \mathcal{D}_\star^{\beta_2} x(t) + [x(t)]^2 = t^6 + \frac{6t^{3-\gamma}}{\Gamma(4-\gamma)} + \frac{36t^{6-\beta_1-\beta_2}}{\Gamma(4-\beta_1)\Gamma(4-\beta_2)}, \quad 0 < t < 1,$$

where $2 < \gamma < 3$, $0 < \beta_1 < 1$, and $1 < \beta_2 < 2$ and the initial conditions are $x(0) = 0$, $x^{(1)}(0) = 0$, and $x^{(2)}(0) = 0$. An easy calculation shows that $x(t) = t^3$ is the exact solution.

For this example, we set $\alpha = 1$. By applying the collocation technique based upon new Bessel functions at C1: $(\gamma, \beta_1, \beta_2) = (5/2, 9/10, 3/2)$ and for $N = 3, 4$, the following approximative solutions on $0 \leqslant t \leqslant 1$ are obtained

$$x_{3,1}(t) = 1.0000000000004519163\, t^3 - 5.55112 \times 10^{-17}\, t - 5.55112 \times 10^{-17},$$
$$x_{4,1}(t) = 8.60154 \times 10^{-13}\, t^4 + 0.9999999999992085443\, t^3 - 1.09400 \times 10^{-16}\, t^2 - 9.83260 \times 10^{-17}\, t + 1.79230 \times 10^{-17}.$$

A comparison between our collocation scheme at C1 and the method of shifted Jacobi operational matrix (SJOM) [6] with $N = 24$ is made in Table 4. Besides the cases C1 and C2: $(\gamma, \beta_1, \beta_2) = (2.000001, 0.000009, 1.000001)$, the following values of $(\gamma, \beta_1, \beta_2)$ are used in Table 5 for comparison purposes

C3: $(2.99, 0.99, 1.99)$, C4: $(2.75, 0.75, 1.75)$, C5: $(2.9999, 0.9999, 1.9999)$.

Table 4. Comparison of L_∞ error norms for $\gamma = 5/2, \beta_1 = 9/10, \beta_2 = 3/2$ in test Problem 8.

	New Bessel		**SJOM ($N = 24$) [6]**			
	$N = 3$	$N = 4$	$\alpha, \beta = 0$	$\alpha, \beta = \frac{1}{2}$	$\alpha, \beta = 1$	$\alpha, \beta = \frac{3}{2}$
L_∞	4.51805_{-13}	6.85082_{-14}	3.37_{-05}	3.50_{-05}	3.39_{-05}	3.15_{-05}

Table 5. Comparison of L_∞ error norms for various $(\gamma, \beta_1, \beta_2)$ in test Problem 8.

	New Bessel		**SJOM ($\alpha, \beta = \frac{3}{2}$) [6]**			
Case	$N = 3$	$N = 4$	$N = 4$	$N = 8$	$N = 16$	$N = 24$
C2	4.14718_{-12}	3.82214_{-15}	1.47_{-09}	2.43_{-10}	2.62_{-11}	6.29_{-12}
C3	0	1.87623_{-16}	1.85_{-04}	5.32_{-05}	3.50_{-05}	1.95_{-05}
C4	0	3.48011_{-14}	2.02_{-03}	5.93_{-04}	2.40_{-04}	1.06_{-04}
C5	0	4.49186_{-16}	1.91_{-06}	5.46_{-07}	3.67_{-07}	2.06_{-07}

Looking at Tables 4 and 5 reveals that our numerical solutions obtained via novel Bessel-collocation method are in excellent agreement with the corresponding exact solutions. Moreover, our proposed scheme is superior compared to the SJOM.

Problem 9. *We consider the fractional relaxation-oscillation equation [5,6]*

$$\mathcal{D}_\star^\gamma x(t) + x(t) = 0, \quad 0 < \gamma < 2,$$

with the initial condition $x(0) = 1$. If $\gamma > 1$ we also have $x^{(1)}(0) = 0$. The exact solution in terms of Mittag–Leffler function is given by $x(t) = E_\gamma(-t^\gamma)$. Here, $E_\gamma(z) = \sum_{k=0}^{\infty} \frac{z^k}{\Gamma(1+\gamma k)}$.

First, we consider $\gamma = 85/100$ and set α equals to γ. We get the approximated solution $x_{N,\alpha}(t)$ using $N = 8$ terms on $[0, 1]$ as follows

$$x_{8,\frac{85}{100}}(t) = 0.0972690897737097\, t^{\frac{17}{5}} - 0.0264284381134049\, t^{\frac{17}{4}} + 0.647219778384659\, t^{\frac{17}{10}} - 1.05749619232596\, t^{\frac{17}{20}} + 0.00525953072336809\, t^{\frac{51}{10}} - 0.284023703239741\, t^{\frac{51}{20}} - 0.000568791172776014\, t^{\frac{119}{20}} + 1.0.$$

In Table 6, we calculate the maximum absolute errors using $N = 8$ and $N = 10$. In addition, a comparison is done in this table with the results obtained via SJOM [6]. Looking at Table 6 one can find that the achievement of good approximations to the exact solution is possible using only a few terms of fractional Bessel polynomials.

Table 6. Comparison of L_∞ error norms for $\gamma, \alpha = 85/100$ in test Problem 9.

	New Bessel		SJOM ($N = 32$) [6]			
	$N = 8$	$N = 10$	$\alpha, \beta = -\frac{1}{2}$	$\alpha, \beta = 0$	$\alpha, \beta = \frac{1}{2}$	$\alpha, \beta = 1$
L_∞	1.01411_{-06}	6.16222_{-09}	5.2_{-04}	8.1_{-05}	1.2_{-04}	2.3_{-04}

In the next experiments, we investigate the impact of varying γ on the maximum absolute errors while $N = 10$ is fixed. Table 7 presents the L_∞ errors for $\gamma = 0.2, 0.4, 0.6, 0.8$ as well as $\gamma = 1.2, 1.4, 1.6, 1.8$. In all cases, we exploit $\alpha = \gamma$. Comparisons with existing approximation techniques based on operational matrix of fractional derivatives via B-spline functions [9] and shifted Jacobi functions [6] are also carried out in Table 7.

Table 7. Comparison of L_∞ error norms for $N = 10$ and various γ in test Problem 9.

	New Bessel	B-Spline [9]	SJOM ($N = 10$) [6]			
$\gamma = \alpha$	$N = 10$	$J = 8$	$\alpha, \beta = -\frac{1}{2}$	$\alpha, \beta = 0$	$\alpha, \beta = \frac{1}{2}$	$\alpha, \beta = 1$
0.2	6.71097_{-07}	5.3_{-03}	0.2544	0.1684	0.1824	0.1907
0.4	1.05544_{-06}	1.9_{-03}	0.1002	0.0363	0.0489	0.0617
0.6	3.38325_{-07}	1.5_{-03}	0.0314	0.0100	0.0158	0.0202
0.8	1.64178_{-08}	1.0_{-03}	0.0069	0.0018	0.0034	0.0045
1.2	1.53515_{-12}	2.5_{-03}	0.0222	0.0046	0.0046	0.0061
1.4	8.04611_{-15}	2.4_{-03}	0.0085	0.0014	0.0026	0.0041
1.6	6.41447_{-16}	—	0.0031	3.8_{-04}	0.0016	0.0029
1.8	1.17134_{-15}	—	0.0012	7.3_{-05}	9.0_{-04}	0.0016

Problem 10. *In the last case example, let us consider the following singular fractional Lane-Emden type equation [28,29]*

$$\begin{cases} \mathcal{D}_\star^\gamma x(t) + \frac{k}{t^{\gamma-\beta_1}} \mathcal{D}_\star^{\beta_1} x(t) + \frac{1}{t^{\gamma-2}} x(t) = g(t), \quad 0 < t \leqslant 1, \\ x(0) = 0, \quad x^{(1)}(0) = 0, \end{cases}$$

where $1 < \gamma \leqslant 2$, $0 < \beta_1 \leqslant 1$, $k \geqslant 0$, *and*

$$g(t) = t^{2-\gamma} \left(6t \left(\frac{t^2}{6} + \frac{\Gamma(4-\beta_1) + k\Gamma(4-\gamma)}{\Gamma(4-\beta_1)\Gamma(4-\gamma)} \right) - 2 \left(\frac{t^2}{2} + \frac{\Gamma(3-\beta_1) + k\Gamma(3-\gamma)}{\Gamma(3-\beta_1)\Gamma(3-\gamma)} \right) \right).$$

The exact solution is $x(t) = t^3 - t^2$.

To proceed, we take $\gamma = 3/2$, $\beta_1 = 1/2$, and $k = 2$. Using the collocation points $t_j = 0.001 + j/N$ for $j = 0, 1, \ldots, N$ and with $N = 3, 4$, the following approximation solutions are obtained by the Bessel-collocation procedure

$$\begin{aligned} x_{3,1}(t) &= 1.0\,t^3 - 1.0\,t^2 + 6.23678 \times 10^{-16}\,t + 3.40637 \times 10^{-108}, \\ x_{4,1}(t) &= -2.93410 \times 10^{-15}\,t^4 + 1.0\,t^3 - 1.0\,t^2 + 1.28989 \times 10^{-15}\,t + 3.06104 \times 10^{-108}. \end{aligned}$$

Obviously, these approximations are accurate up to machine epsilon. Table 8 reports the comparison of the absolute errors evaluated at some points $t \in [0, 1]$ obtained by the Bessel-collocation method. For comparison, the results obtained by the collocation method (CM) [29] and the reproducing kernel method (RKM) [28] are also shown in Table 8. The comparisons show that the proposed method is considerably more accurate than other methods.

Table 8. Comparison of absolute errors for $\gamma = 3/2, \beta_1 = 1/2$ in test Problem 10.

	New Bessel		CM [29]		RKM [28]	
t	$N = 3$	$N = 4$	$N = 5$	$N = 10$	$N = 5$	$N = 10$
0.25	8.25341_{-17}	1.32991_{-16}	1.3345_{-03}	1.3232_{-05}	8.7370_{-04}	8.4636_{-06}
0.50	6.41715_{-17}	1.23861_{-16}	1.5000_{-03}	2.6000_{-05}	9.9000_{-04}	2.9000_{-06}
0.75	1.37231_{-17}	1.21568_{-16}	5.0673_{-03}	1.5634_{-06}	7.6702_{-04}	8.5754_{-06}
1.00	3.40637_{-108}	3.75169_{-108}	3.6339_{-03}	4.1443_{-05}	5.4736_{-04}	5.4345_{-06}

6. Conclusions

A practical matrix approach based on novel (orthogonal) Bessel polynomials is presented to solve multi-order fractional-order differential equations (MOFDEs). Using the matrix representations of the generalized Bessel polynomials and their derivatives with the aid of collocation points, the scheme transforms MOFDEs to a fundamental matrix equation, which corresponds to a system of (non)linear algebraic equations. To assess the efficiency and accuracy of the presented technique, several numerical examples with initial and boundary conditions are investigated. Comparisons with the exact solutions and with various alternative numerical simulations and experimental measurements have also been made. Based on the experiments, it is found that the numerical approximations are in an excellent agreement, which demonstrate the reliability and the great potential of the presented technique.

Author Contributions: Conceptualization, M.I. and C.C.; methodology, M.I. and C.C.; software, M.I.; validation, M.I. and C.C.; formal analysis, M.I. and C.C.; investigation, M.I. and C.C.; writing—original draft preparation, M.I. ; writing—review and editing, M.I. and C.C. All authors have read and agreed to the published version of the manuscript.

Funding: This research received no external funding.

Conflicts of Interest: The authors declare no conflict of interest.

References

1. Podlubny, I. *Fractional Differential Equations*; Academic Press: New York, NY, USA, 1999.
2. El-Sayed, A.M.A.; El-Mesiry, E.M.; El-Saka, H.A.A. Numerical solution for multi-term fractional (arbitrary) orders differential equations. *Comput. Appl. Math.* **2004**, *23*, 33–54. [CrossRef]
3. Edwards, J.T.; Ford, N.J.; Simpson, A.C. The numerical solution of linear multi-term fractional differential equations: systems of equations. *J. Comput. Appl. Math.* **2002**, *148*, 401–418. [CrossRef]
4. El-Sayed, A.M.A.; El-Kalla, I.L.; Ziada, E.A.A. Analytical and numerical solutions of multi-term nonlinear fractional order differential equations. *Appl. Numer. Math.* **2010**, *60*, 788–797 [CrossRef]
5. Saadatmandi, A.; Dehghan, M. A new operational matrix for solving fractional-order differential equations. *Comput. Math. Appl.* **2010**, *59*, 1326–1336. [CrossRef]
6. ABhrawy, H.; Taha, T.M.; Machado, J.A.T. A review of operational matrices and spectral techniques for fractional calculus. *Nonlinear Dyn.* **2015**, *81*, 1023–1052. [CrossRef]
7. Talaei, Y.; Asgari, M. An operational matrix based on Chelyshkov polynomials for solving multi-order fractional differential equations. *Neural Comput. Appl.* **2018**, *30*, 1369–1376. [CrossRef]
8. Izadi, M. A comparative study of two Legendre-collocation schemes applied to fractional logistic equation. *Int. J. Appl. Comput. Math.* **2020**, *6*, 71. [CrossRef]
9. Lakestani, M.; Dehghan, M.; Irandoust-pakchin, S. The construction of operational matrix of fractional derivatives using B-spline functions, Commun. *Nonlinear Sci. Numer. Simul.* **2012**, *17*, 1149–1162. [CrossRef]
10. Damarla, S.K.; Kundu, M. Numerical solution of multi-order fractional differential equations using generalized triangular function operational matrices. *Appl. Math. Comput.* **2015**, *263*, 189–203. [CrossRef]
11. Krall, H.L.; Frink, O. A new class of orthogonal polynomials: The Bessel polynomials. *Trans. Am. Math. Soc.* **1949**, *65*, 100–115. [CrossRef]
12. Han, H.; Seo, S. Combinatorial proofs of inverse relations and log-concavity for Bessel numbers. *Eur. J. Combin.* **2008**, *29*, 1544–1554. [CrossRef]
13. Yang, S.L.; Qiao, Z.K. The Bessel numbers and Bessel matrices. *J. Math. Res. Exp.* **2011**, *31*, 627–636.

14. Ismail, M.E.H.; Kelker, D.H. The Bessel polynomials and the student t distribution. *SIAM J. Math. Anal.* **1976**, *7*, 82–91. [CrossRef]
15. Bochner, S. Über Sturm-Liouvillesche polynomsysteme. *Math. Zeit.* **1929**, *29*, 730–736. [CrossRef]
16. Grosswald, E. On some algebraic properties of the Bessel polynomials. *Trans. Am. Math. Soc.* **1951**, *71*, 197–210. [CrossRef]
17. Grosswald, E. *Bessel Polynomials, Lecture Notes in Math. Vol. 698*; Springer: Berlin, Germany, 1978.
18. Delgado, J.; Orera, H.; Pe na, J.M. Accurate algorithms for Bessel matrices. *J. Sci. Comput.* **2019**, *80*, 1264–1278. [CrossRef]
19. Dehestani, H.; Ordokhani, Y.; Razzaghi, M. Fractional-order Bessel functions with various applications. *Appl. Math.* **2019**, *64*, 637–662. [CrossRef]
20. Odibat, Z.; Shawagfeh, N.T. Generalized Taylor's formula. *Appl. Math. Comput.* **2007**, *186*, 286–293. [CrossRef]
21. Kazem, S.; Abbasbandy, S.; Kumar, S. Fractional-order Legendre functions for solving fractional-order differential equations. *Appl. Math. Model.* **2013**, *37*, 5498–5510. [CrossRef]
22. Rehman, M.U.; Khan, R.A. A numerical method for solving boundary value problems for fractional differential equations. *Appl. Math. Model.* **2012**, *36*, 894–907. [CrossRef]
23. Firoozjaee, M.A.; Yousefi, S.A.; Jafari, H.; Baleanu, D. On a numerical approach to solve multi order fractional differential equations with initial/boundary conditions. *J. Comput. Nonlinear Dyn.* **2015**, *10*, 061025. [CrossRef]
24. Izadi, M.; Negar, M.R. Local discontinuous Galerkin approximations to fractional Bagley-Torvik equation. *Math. Meth. Appl. Sci.* **2020**, *43*, 4978–4813. [CrossRef]
25. Wei, H.M.; Zhong, X.C.; Huang, Q.A. Uniqueness and approximation of solution for fractional Bagley-Torvik equations with variable coefficients. *Int. J. Comput. Math.* **2016**, *94*, 1541–1561. [CrossRef]
26. Izadi, M. Fractional polynomial approximations to the solution of fractional Riccati equation. *Punjab Univ. J. Math.* **2019**, *51*, 123–141.
27. Izadi, M. *Comparison of Various Fractional Basis Functions for Solving Fractional-Order Logistic Population Model*; Facta Univ. Ser. Math. Inform; University of Niš: Niš, Serbia, 2020.
28. Akgul, A.; Inc, M.; Karatas, E.; Baleanu, D. Numerical solutions of fractional differential equations of Lane-Emden type by an accurate technique. *Adv. Differ. Equ.* **2015**, *2015*, 220. [CrossRef]
29. Mechee, M.S.; Senu, N. Numerical study of fractional differential equations of Lane-Emden type by method of collocation. *Appl. Math.* **2012**, *3*, 851–856. [CrossRef]

MDPI
St. Alban-Anlage 66
4052 Basel
Switzerland
Tel. +41 61 683 77 34
Fax +41 61 302 89 18
www.mdpi.com

Symmetry Editorial Office
E-mail: symmetry@mdpi.com
www.mdpi.com/journal/symmetry

www.ingramcontent.com/pod-product-compliance
Lightning Source LLC
LaVergne TN
LVHW070619170726
843515LV00004B/123